Generative AI in Writing Education

This book provides a theoretical framework to allow educators, researchers, and policymakers to better understand computer-generated writing and the policy and pedagogical implications of generative AI.

Generative AI, such as ChatGPT and Gemini, has substantially disrupted educational spaces, forcing educators, policymakers, and other stakeholders to reconsider writing and how it should be used in education. Responding to this disruption, this book provides technically sound guidance on how various stakeholders should engage with generative AI. After providing a foundational and technical discussion of the technology, this book directly addresses the educational context. Informed by theories of learning and knowledge transfer and utilizing rhetorical theories of writing, this book assesses the impact of AI on student learning, student performance, and academic honesty and integrity. In doing so, the book outlines how generative AI can be both a help and a hindrance for students, enabling readers to craft informed and meaningful policies and successfully integrate AI in the composition classroom.

This book will be of interest to scholars in the fields of Rhetoric and Composition, Technical Writing, Communication Studies, Linguistics, and TESOL, as well as to Education and Machine Learning policymakers, program directors, and researchers.

Dylan Medina is an Acting Assistant Professor in the Department of English at the University of Washington, USA. He is also the Director of Software Engineering at gotLearning.

Routledge Research in Writing Studies

Writing, Imitation, and Performance
Insights from Neuroscience Research
Irene L. Clark

Emotional Value in the Composition Classroom
Self, Agency, and Neuroplasticity
Ryan Crawford

Dialogic Editing in Academic and Professional Writing
Engaging the Trace of the Other
Edited by Özüm Üçok-Sayrak, Janie Harden Fritz and Kristen Lynn Majocha

A Multidisciplinary Exploration into Flow in Writing
Deborah F. Rossen-Knill, Katherine Schaefer, Matthew W. Bayne, Whitney Gegg-Harrison, Dev Crasta, and Alessandra R. Dimauro

Generative AI in the English Composition Classroom
Practical and Adaptable Strategies
Daniel Plate, Elizabeth Melick, James L. Hutson, and Susan Edele

Generative AI in Writing Education
Policy and Pedagogical Implications
Dylan Medina

For more information about this series, please visit: https://www.routledge.com/Routledge-Research-in-Writing-Studies/book-series/RRWS

Generative AI in Writing Education

Policy and Pedagogical Implications

Dylan Medina

NEW YORK AND LONDON

First published 2025
by Routledge
605 Third Avenue, New York, NY 10158

and by Routledge
4 Park Square, Milton Park, Abingdon, Oxon, OX14 4RN

Routledge is an imprint of the Taylor & Francis Group, an informa business

ISBN: 9781032792491 (hbk)
ISBN: 9781032797328 (pbk)
ISBN: 9781003493563 (ebk)

DOI: 10.4324/9781003493563

Typeset in Times New Roman
by codeMantra

Contents

Introduction

In late November 2022, OpenAI released the famous ChatGPT, a chatbot that could produce text that was very difficult to distinguish from natural human writing. ChatGPT implements a natural language generation pipeline that includes a GPT style large language model (LLM). This technology was significant because unlike with prior chatbots and natural language generation systems, users of ChatGPT quickly found that it could be used to automate writing tasks sufficiently well, and as a result, people turned to it to avoid having to do the difficult and often boring work of writing. In education, this posed a significant problem because writing is frequently used for assessment and grading, and ChatGPT and other LLM-based chatbots could easily be used by students to complete these writing tasks. Educators and scholars have responded and offered a wide array of how we might respond. While antiplagiarism countermeasures have been a major focus of the discussion, even more useful responses have focused on how we might incorporate this technology into our writing practice and instruction. While there are risks and benefits to the advent of generative AI systems, they are likely here to stay, and so there is a considerable opportunity for educators, and Writing Studies and Communication scholars, in particular, to develop a robust landscape of theory that positions these systems in our larger understanding of rhetoric and expansive pedagogical frameworks for incorporating natural language generation into our writing instruction. This book is a tentative attempt to engage in this work.

While ChatGPT has created a great deal of shock in the community, it is really the next step in a long history of work on natural language processing and natural language generation. Since we have had computers, we've wanted to build computation systems that could interact with natural human language in meaningful ways. However, this has proven to be a significant challenge largely because language is a system that is sufficiently complex and changeable that it is very difficult to create rules that effectively model it. Natural language processing—or the use of algorithms to extract information from natural human language—made advances more quickly than natural language generation—or the use of algorithms to generate natural language. The benefit of natural language processing is that it makes it more realistic to make general claims about language based on data sets that are sufficiently large that they might be considered representative of language, in general. Natural language processing tools allow researchers to collect massive corpora of human language and extract interesting information from those collections of texts. Natural language generation is more difficult because while it is not terribly difficult to write instructions that a computer could follow to construct grammatically correct human language, it is very difficult to write instructions that a computer could follow to construct natural human language that responds to rhetorical situations in appropriate ways. As will be discussed in the second and third chapters, the set of situations that bring language into being and the various appropriate responses to any of those situations are so large that it is quite difficult to create instructions for all of them. This book will discuss at length the ways that technology has advanced to come closer to being able to accomplish natural language generation.

I have been working with natural language processing and natural language generation tools over a decade. One of the earlier projects I worked on used concordances and analyzed the position of language used by radio personalities in relation to dominant American English. I also developed tools to identify lexical density, sentiment, and off topic sentences. During the winter holiday in 2022, I started working with OpenAI's application programming interface (API), which allowed me to write programs that used OpenAI's GPT-3.5 model to generate text. An API like that offered by OpenAI is an internet resource that can be accessed by URL via the internet. The API provides a particular protocol, and the user can make requests to the URL using the protocol, and the API will provide predictable responses. At the end of the break, I had a toy that could generate feedback on student writing. I could send chunks of writing in a particular format, and OpenAI's API would respond with generated text that is feedback to the writing I sent. Later, my work with gotLearning allowed me to build that into a tool that we released publicly. As I experimented with this tool, I discovered that the GPT-3.5 model could generate feedback that accurately identified features of the input writing. In other words, if the input writing lacked sufficient evidence supporting a claim, the system can identify that. However, the system cannot account for the particulars of the writer. This reveals a great deal about how GPT models work. While students are each unique, there are clear patterns to the features in their writing that a teacher might want to comment on. Many students, for instance, do not offer enough evidence, or do too little to explain how the evidence supports a claim, and so on. The fact that the system using the GPT-3.5 model could accurately identify and respond to these features suggests that the corpus on which the model was trained contained enough information about this pattern in writing that it is included in the model. After releasing this tool, I decided to start working on this project and hopefully make some contribution to the emerging conversation on generative AI.

I have used generative AI during the process of writing this book. I have used it in three distinct ways. I spent considerable time interacting with ChatGPT-3.5. My goal in this was to better understand what kind of text it could generate and where it would fail. A good activity to understand it is to try to get the system to hallucinate or create text that is wrong for some reason. I learned a great deal about the system from a user's perspective. While I do not include any of the text generated in these interactions in this book, I did use the technology to experiment. Second, I used OpenAI's API to experiment with the various configurations and prompt design. These illustrate the way that, for instance, a configuration called "temperature" can introduce artificial randomness in the predictions. Likewise, through prompt engineering, or the way in which one designs a prompt, reveals a great deal about how these LLMs generate text and position that text in discourse. For instance, if I prompt the system to respond as if it were a genius with an IQ of 1,000, the system matches the prompt with fiction and is more likely be fictional or fantastical in its response. Again, none of these interactions are included in this book but were important in informing my perspective. Finally, I did explicitly include text generated by both ChatGPT-4o and Gemini 1.5 as an object for analysis in this book. The generated text is clearly cited and referenced. I use ChatGPT and Gemini to generate texts in response to particular prompts and analyze the generated texts to illustrate the way that we might think about these technologies in the context of rhetorical theory.

As I conclude this introduction, I'd like to emphasize that this book is an early effort to position generative AI in rhetorical theory and writing pedagogy. One of my primary goals is to open the blackbox that is generative AI so that our future conversions can be informed by a more expansive understanding of what the technology does and how it works. I hope that future conversations can extend our understandings of generative AI in the context of theory and pedagogy in the fields of Writing Studies, Communication Studies, Education, and other related fields.

Acknowledgments

This book has been possible thanks to a great deal of support and guidance that I have received over a number of years from my colleagues, friends, and family. First, I'd like to express thanks to my mentors and guides who helped me expand my research interests beyond Writing Studies into Digital Humanities. In particular, I'd like to thank Rob Weller for mentoring me as I began to learn computer programming and for giving me the space to explore various technologies while I worked as an assistant computer specialist for the University of Washington's English Department. Our conversations during this project have also helped me organize my lines of argument. I would also like to thank Shawn Rider for our numerous conversations about Python, natural language processing, and generative AI. During this project, he shared a number of resources that were useful in seeing into the blackbox of these technologies. I would also like to thank Anis Bawarshi for our conversations early on in the project. He helped me find the position that I would take up and pointed out the work that Sid Dobrin was doing in developing pedagogy that accommodates students having access to generative AI. Additionally, since this is my first book-length project, I would particularly like to thank Anis for his guidance in finding an approach.

I would also like to thank the numerous educators who have helped me gain access to the field of software engineering. My graduate degree is in Writing Studies, and computer programming had been a side activity that served my main work of writing instruction. As a faculty member at Seattle University, the university funded my ongoing education, and there I earned a Computer Science certificate, which was designed as a transition pathway from humanities and social science degrees into Computer Science. This program not only taught how to write code, but more importantly, it served as a crucial gateway into the discourse and dispositions of the field. The program taught us computational thinking and effectively prepared us to enter into the field of Computer Science and software engineering.

I am also grateful for the many conversations that I had with my colleagues and students in the Digital Technologies and Cultures program at Seattle University. This now-closed program helped prepare post-traditional students to be successful in careers that bridged the gap between the humanities and STEM. In particular, the conversations I had with my colleagues about social media bots were particularly informative. Communications researchers have long studied the rhetoric of bots on social media and understanding concerns around their efficacy and ethics have contributed significantly to how I think about generative AI, since the technologies are related (most of the social media bots of the late 2010s used intent-based ML as I describe in Chapter 3). I also want to thank my students, most of whom were coming from humanities backgrounds. Over the years as I developed and revised courses on Python programming and Concepts in Computing, I learned about how to make understanding computers more accessible.

Finally, I am grateful to my spouse, Émilie Medina, for her patience, support, and understanding. Émilie has been there to encourage me throughout the entire process, and for that I am grateful.

1 How I Learned to Stop Worrying and Love AI

1.1 Initial Reactions

In late 2023, OpenAI released ChatGPT-3.5 to the public sending shockwaves through public and academic discourses alike. There has been a long history of what Amy Orben (2020) calls "technology panics" that seem to appear with the rise of every technology that has the potential to change the way we live. It is true that novel technologies change our world experience, which generally include risks and benefits. Even before OpenAI released the previous version of their famous chat app (ChatGPT-3.0), *The Atlantic* published a boldly titled article, "The College Essay is Dead" (Marche, 2022). The article begins by highlighting the challenges teachers will face with the expanded opportunities for plagiarism made possible by generative artificial intelligence (GenAI)[1] and laments the slow decline of the humanities and literate culture. Marche references Mark Zuckerberg's lack of literacy of "17th century pamphlets" and Elon Musk's famous rejection of "books" (Marche, 2022). In no way am I setting up a defense of Musk or Zuckerberg, but the point is panicking over perceived threats to literacy is not new and perhaps not particularly useful. Likewise, *The Chronicle for Higher Education* has published many opinion pieces on GenAI since the end of 2023 (*Advice | Are We Asking the Wrong Questions about ChatGPT?*, 2024; *Advice | The Case for Slow-Walking Our Use of Generative AI,* 2024; *Artificial Intelligence and the* Significance Crisis, 2024; *Opinion | How Will Artificial Intelligence Change Higher Ed?*, 2023). Much of this discourse boils down to a collective anxiety over a threat to the place of writing in education, and more generally questions about the purpose of student work. If GenAI can produce work that is often indistinguishable from the work produced by a student, there is a real potential problem because the point of work in school is as much in *doing the work* rather than submitting some product. I suggest that one potential benefit of GenAI is the way it forces educators, administrators, and policymakers to reconsider the place of writing in their classes and institutions. The fourth chapter will address this question.

GenAI does pose a particular threat to student learning in the sense that it can be used by students to avoid doing the important work of learning.

DOI: 10.4324/9781003493563-1

However, writing is one of the main tools used for gatekeeping and ranking students, and one might see its central role in the maintenance of the educational-industrial complex. Perhaps some of the panic over GenAI has to do with its disruptive potential to that system. Nearly all academic fields at some point use writing as one of the ways they assess student performance. Classes in the Humanities and the Social Sciences widely use short-answer and essay writing to assess both students' uptake of the course material and students' ability to perform various writing prescribed writing moves. Likewise, many STEM (Science, Technology, Engineering, and Mathematics) courses also use writing in some form, from lab and research reports to mathematical proofs, as part of their assessment. In addition to assessment within particular classes in academic institutions, admission processes frequently use writing to filter people who have applied. Writing genres like personal statements, entrance essays, and writing tests are often used as gatekeeping instruments to grant or restrict access to the institutions. Writing is so universally recognized as a crucial skill for students in higher education that despite critiques from many scholars in Writing Studies, First-Year Composition course is often one of the few mandatory courses that all students must take during college regardless of their field of study (Downs & Wardle, 2007; Miller, 1991; Wardle, 2009). I do appreciate the emphasis placed on writing in our educational systems, and the conversation regarding the role of a mandatory FYC course is worth continuing now that so many students are using GenAI, but I will not be discussing that at any length here. Instead, I will focus this discussion on writing and GenAI on how the technology was built, how it works, and how we might be able to accommodate it into our pedagogy where it has significant potential to enhance learning.

While I think the concerns over plagiarism are valid, I will spend very little time addressing them and will focus primarily on intellectual property in the fourth chapter. It is worth noting that while plagiarism is a concern, it has always been possible to game writing assessment. Students have always been able to hire tutors and ghostwriters, and since the advent of the internet, it has become even easier to purchase a paper from a paper mill. The point being, plagiarism has always been possible, but it was costly and allowed privilege to remain comfortably in the hands of already privileged students. A consideration that is worth further discussion is the way that GenAI threatens to expand access to cheating. To use this technology, one needs an internet-connected device and a free account. There are still obstacles that prevent equitable access to GenAI, but I suspect they are less than expensive tutors or paper writers. In later chapters, I analyze some text generated by the cutting-edge public GenAI chat systems, and I was able to do so without paying anything. With this technology becoming so accessible, scholars, educators, administrators, and companies have offered many solutions to police use of GenAI. The most useful of these solutions encourage educators to become more familiar with their students' writing across multiple drafts so they can see when a student becomes over-reliant on GenAI. They also encourage educators to assign work that privileges student creativity and knowledge production, both of which are difficult for GenAI to do.

Rethinking how we assign student work and inviting—and teaching—students to use GenAI responsibly seem to be the most reasonable responses to the disruption caused by the technology. I want to foreground Cress and Kimmerle's (2023) "Co-constructing knowledge with generative AI tools: Reflections from a CSCL perspective" and Sid Dobrin's (2023) book, *AI and Writing,* as exemplary work that builds on accurate perspectives on GenAI and provides useful pedagogical insights and teaching strategies. As these scholars suggest, we can teach students to use GenAI as a tool in their writing process. I would add that prior to further research into how we should position GenAI in our fields, we can immediately address some of the concerns by adjusting what we want students to do with writing. We might first spend more time with the student during their writing process, and second, we might focus our assignments more on the production of knowledge rather than information and recall. We already have an entire movement in Writing Studies—the "writing to learn" movement—that treats writing instruction as an opportunity for students to create knowledge. The approach of spending more time with students during their writing process has the potential because it becomes easier to identify writing that is not the students' own. When writing is assigned as a terminal assessment used to determine if students have learned some material, it is easy for students to use GenAI without being detected. However, if teachers work with their students through an ongoing process, the teachers become familiar with the student and their voice, and at that point, it becomes easy for them to identify GenAI. In this case, even if the students use GenAI to help them along the way, the interactive process with the teacher involved means that the student is learning, which is the goal anyway. If writing is treated as a series of interactions, which I argue elsewhere benefits learning because learning occurs in these moment-to-moment interactions—or "microtransfers," then teachers and students can use GenAI as a tool to further their learning (Medina, 2018). Additionally, focusing on knowledge production rather than knowledge retrieval in writing tasks makes it difficult for students to use GenAI because this technology excels at knowledge retrieval and is incapable of generating new ideas. The reason for this will become clear in the next chapter. When we ask students to write essays and summaries to assess whether they have done the reading and have retrieved the desired ideas from the course materials, we are asking them to do something at which GenAI is quite effective. Since the large language models (LLMs) at the heart of GenAI systems were constructed by extracting features that exist in their training corpora, and since these corpora most certainly contain both the materials that teachers ask students to write about and samples of student writing, they are very good at reconstructing the writing on which they were trained and thus the writing teachers are asking students to do.

While there are changes that we can make to address the more major concerns over GenAI, I think that educators, administrators, policymakers, and scholars whose fields are being disrupted by this technology will benefit most by resolving some of the misconceptions about GenAI. This book aims to do this through three main goals. First and foremost, I set out to build some foundational understanding into how researchers solved a particular problem to develop LLMs and GenAI, how the technology came into being, and how it works now. From this foundation, I will position GenAI in the context of some rhetorical theory with the hopes of beginning an extended conversation on how it affects language, writing, and rhetoric. Finally, I will discuss some starting points for pedagogy and research that incorporate existing scholarship and points to potential paths forward. I very much see this project as a tentative starting point for conversations around GenAI, writing with it, and how it helps us understand rhetoric and education. The remainder of this chapter focuses on the MLA-CCCC's Joint Task Force on Generative AI's working paper because it carries considerable weight and some limitations of the paper served as impetus for this project.

In the second chapter, I will go in depth into the heuristics used by Computer Science to develop solutions to human problems. I will also discuss the definition of "intelligence" as it pertains to artificial intelligence. This discussion focuses less on GenAI but it provides critical background into how software and hardware are developed. By understanding the discourses involved in producing technology, we will be in a better position to understand and comment on that technology itself. Further, "intelligence" itself is a challenging concept, and differentiating between how computers process information and how humans process information is important. While human intelligence is particularly complex, artificial intelligence that uses machine-learning techniques is perhaps more straightforward because ultimately it is the complex assembly of a lot of mathematical calculations.

In the third chapter, I turn to the story of how Computer Scientists, Data Scientists, and Computational Linguists have solved what I am calling the "language problem," or the challenge of building systems that allow computers to interact with human language. This is a difficult problem, and the solutions will not only help us better understand GenAI but also how human language might be modeled by computer systems. The purpose of this chapter is to delve more profoundly into the technical aspects of GenAI. The scholarship on GenAI outside of Computational Linguistics and Computer Science on the topic glosses over the technology. Frequently little more is provided than a quick reference to Stephen Wolfram's (2023) excellent blog post and an explanation that GenAI generates text word by word by predicting the next most probable word in a sequence. While this explanation of GenAI is correct, I suggest that people making policy on GenAI and educators who are encountering students using it should have more than a cursory understanding of how the technology works. In this chapter, I will discuss how language has been modeled with weighted matrices and algorithms so that computers can process it. Computers require these mathematical models because computers can only work with numbers. They cannot directly compute the complexities and nuances that are part of natural language, and so those complexities must be encoded. I will discuss the various approaches being taken. Finally, this chapter will discuss the variety of LLMs and GenAI systems suggesting that we cannot think of them all as a single technology or innovation.

In the third chapter, I will review the literature on GenAI in the field of Writing Studies. Much of this literature grapples with GenAI in useful ways and provides excellent pedagogical recommendations; however, there is considerable room to say more about the GenAI systems, the LLMs' representation of language, and their relationship to rhetorical theory and pedagogy. Moreover, one of the recurring metaphors in the literature is that of "collaboration" (Anderson, 2023). It is understandable why we might consider ourselves to be collaborating with GenAI, but this is a highly inaccurate metaphor and even problematic. This is particularly true when we use this metaphor with students whose language practices differ from those expected by the university. For underrepresented students from non-dominant discourses and languages, GenAI particularly threatens to erase their voices, and the collaboration metaphor amplifies this erasure, and so I will push back strongly against this metaphor. Instead, I will frame LLMs and GenAI systems in conversation with rhetorical genre studies (RGS). RGS is particularly effective for discussing this technology because it theorizes the "regularity and chaos" involved in writing genres (Devitt, 2004, p. 156). LLMs are models of the regularity or conventions of language and can be used in negotiations that occur within a genre. I will also position GenAI in the context of post-humanist theories of rhetoric that see language as a function of a complex system that includes but extends beyond the individual language user. These theories allow us to consider the complex human, material, and technological assemblages that interact forming ecologies, systems of complexity, and flows that produce rhetorical actions (Dobrin, 2011; Edbauer, 2005; Hawk, 2007). GenAI that we see is, in fact, more than a model of language, but it is instead a complex system, and when used in writing, that system interacts with the audience, the rhetorical situation, and the human writer. This theory is particularly well suited to trace the interactions that exist there.

In the fourth chapter, I will make the turn to pedagogy and research implications. I will begin by discussing plagiarism, but I limit that conversation to the intellectual property component of plagiarism because the academic integrity aspect of plagiarism has been a major part of existing scholarship. This chapter will also address a particular threat that GenAI poses to translingual, antiracist, and anticolonial perspectives in Writing Studies. LLMs are trained on mostly what A. Suresh Canagarajah (2006) would call "metropolitan English" (ME)—the dominant English spoken in the United States, the UK, New Zealand, and Australia by people in positions of power and authority in those discourses. As a result, they model that particular English and are particularly adept at generating text that conforms to the standards of ME. Students using this technology might benefit by gaining access to ME, but this could be at the expense of their agency and the translingual potential of their language practices. Even though my expertise in the area of Translingualism is limited to practical implementations in my own teaching, I will raise this topic because I believe that the topic is significantly important, and the threat posed by GenAI to diversity and inclusion is so great. After addressing these significant risks, I will turn to the considerable amount of existing scholarship and recommendations on how we might address GenAI in our pedagogy. This chapter will position some of the more exemplary recommendations alongside the prior technical and theoretical discussions of GenAI, and I will propose several additional strategies, as well. I suggest that GenAI can be incorporated quite effectively into our pedagogy and teaching practices, and the main challenge educators and policymakers should pursue is how can we foster learning and effective writing with GenAI as one of many technologies that students and writers can use in their writing. In addition to discussing pedagogical implications, Chapter 4 will also address research implications. While GenAI is the continuation of decades of research into developing algorithms and models that can interact with human language first in syntactic and later in semantic ways, Writing Studies has just begun to consider the implications of this for our understanding of learning and language. There is presently considerable room and need for more research into GenAI and LLMs, and even as I write this, they are continuing to advance. This project began when GPT-3.5 was one of the most innovative language models, and as I approach the end, Google Gemini, GPT-4o, and others have emerged onto the scene as rivals that perform significantly better than GPT-3.5 did. Moreover, at the beginning of 2024, researchers introduced new models that use 1-bit values within the weighted matrices rather than 16-bit floating point values (the difference between 1, 0, −1, and 0.865342), which presents the potential for greater speed and less hardware requirements. Researchers studying Writing Studies, Education, and Digital Humanities have a significant role to play in how we understand GenAI.

1.2 MLA-CCCC's Joint Task Force on Generative AI

I end this chapter thinking about the MLA-CCCC's Joint Task Force on Generative AI's Working Paper on Generative AI. In late fall of 2023, OpenAI made ChatGPT-3.5 available for public use for free. This chat system uses the GPT-3.5 language model, which is reported to contain 20 billion parameters, meaning 20 billion features of natural language were represented in the model, and for the first time, natural language generation (NLG) became a real possibility. GPT-3.0 had performed well, but this new model within the larger assembly of ChatGPT performs well enough that it could generate texts that could be comfortably passed off as natural language. As I discussed above, this caused a great deal of excitement and anxiety, particularly in fields interested in writing, and within several months, the MLA-CCCC's Joint Task Force on Generative AI released a working paper on the technology. Coming from such authoritative scholarly organizations, this paper represents an equally authoritative view of how the field sees GenAI and its place in education. Working papers, while considerably more tentative than official statements, coming from such academic associations have considerable power in both identifying the dominant currents in the field and grounding the field around particular ideas. For instance, the CCCC's oft-cited statement on "Students' Right to Their Own Language" (2018), originally published in 1974 and reaffirmed several times thereafter, marked a turning point in how writing instructors should treat linguistic diversity. In other words, official statements made on behalf of conferences like the MLA and CCCC's carry significant weight. While working papers are more tentative and perhaps carry less weight, it is important to address some of the issues in their working paper on GenAI.

The working paper begins by summarizing the scholarship on GenAI through the early part of 2023, and they offer a thorough list of risks and benefits associated with using GenAI. The risks presented in the working paper can be summarized into three major categories: risks to critical practice skills, risks associated with equitability and access, and risks associated with academic integrity. The benefits listed in the working paper are much less varied and relate to GenAI's ability to generate human-like natural language texts, for example, and its ability to reveal patterns in language. While the working paper effectively discusses the state of the field and the pedagogical implications of GenAI, it contains some mistakes regarding LLMs and how GenAI systems work. The least significant of these is the authors' explanation of the relationship between LLMs and transformer architecture: "nonprofit OpenAI developed the means to yoke generative pre-trained models to so-called transformer architecture introduced by Google in [2017], thus delivering dramatic increases in performance" (Byrd & Flores, 2023, p. 6). The issue is that the transformer *is* the model, and the LLM used by ChatGPT when the working paper was released was the GPT-3.5 LLM. GPT stands for "generative pre-trained transformer," which means the LLM that is the core of ChatGPT is a transformer model that has been pre-trained for text generation. The major breakthrough in ChatGPT-3.0 and 3.5 is the fact that the models were finally large enough to represent language in a way that allows them to be used to generate what seems like natural language. I will discuss this in greater detail in the next chapter, but a transformer contains what is called "attention mechanisms" and a series of numerical weights. These weights represent features of the data. If we imagine the training data was images instead of text to make it somewhat easier to visualize, the features might be values like the color value or the opacity of pixels in an image. Training involves automatically adjusting these weights based on comparing predicted values with actual values. The breakthrough of transformers is the attention mechanisms which allow the training system to adjust the weights based on the value of a larger span of inputs, whereas previous neural networks could only adjust weights based on one input at a time. If we consider language models, this means that older style neural networks might adjust the weight value by processing one neighbor word at a time within a window of the previous several and following several words. Attention mechanisms allowed each word's weight to be adjusted by the values of all the words in a sentence or pair of sentences, and this can be done in parallel. In short, the breakthrough was architecture that allowed for the processing of more data faster.

Transformer models can be used on diverse types of data and are particularly effective when the relationship between each item is crucial for modeling the item's value. The original transformer-only models described in the paper sponsored by Google focused on automated translation (Vaswani et al., 2017). Transformers work particularly well for modeling natural language because the meaning or significance constituted in natural language is the product of the relationship between the words rather than simply the sum of their individual meanings. Many LLMs, like GPT-3.5, *are* transformer models that have been trained on enormous corpora of natural human language, much of which was scraped from the public internet. Further, GPT-3.5 is not the only LLM that might be used. OpenAI offers several different LLMs themselves with different qualities and many other LLMs exist outside of the OpenAI system. Additionally, ChatGPT is not only an LLM but also it is an entire system that allows the LLM to be used for inference—or generating content—based on some user prompt. The system includes a friendly user interface (UI) and several text-preprocessing and postprocessing steps. The entire system is largely blackboxed—to borrow the concept from Bruno Latour (1999) to describe the hidden nature of the inner working of scientific or technical systems—so that users cannot see exactly what is occurring or how the text is being generated. The important breakthrough that OpenAI made was to train a transformer with a truly staggering number of parameters on massive amounts of natural language and use it in a system that generates content—either text or images.

It is important to note that GenAI are multi-step systems—pipelines is the technical term—and there are several different models and configurations that affect the performance of the system. However, a more significant issue that appears in the working paper is in its explanation of how GenAI uses probability to generate text. The paper explains:

> LLMs work by using statistics and probability to predict what the next character (i.e., letter, punctuation mark, even a blank space) is likely to be in an ongoing sequence, thereby "spelling" words, phrases, and entire sentences and paragraphs. It is not unlike autocomplete, but more powerful… All of the text a model generates is original in the sense that it represents

> combinations of letters and words that generally have no exact match in the training documents, yet the content is also unoriginal in that it is determined by patterns in its training data.
>
> (Byrd & Flores, 2023, p. 6)

When LLMs are used for inference, or to generate text, they do so one "token" at a time, not one "letter" at a time. Intuitively, it would make sense that a token would be a letter because tokens seem like the smallest possible unit of language, but, in fact, a token is a semantic unit—a word or a word part. The distinction is between phonetic and semantic units, and it is important because this kind of NLG uses *units of meaning* as the building blocks. The way that a text is tokenized depends on which tokenizer is used, but the crucial point is that LLMs represent both syntactic and semantic features from language, and so when they are used to generate text, they are, in fact, building meaning.

From our perspective, GenAI may seem a bit like "autocomplete," but the systems are quite different in significant ways (unless, of course, the autocomplete system uses GenAI techniques, which is the case sometimes). It is illustrative to consider LibreOffice's autocomplete system. I am talking about LibreOffice because the suite is similar to Microsoft Office, but it is open source, so we can see how the autocomplete system is coded. Word completions in LibreOffice work because "LibreOffice collects words that you frequently use in the current session. When you later type the first three letters of a collected word, LibreOffice automatically completes the word" (*Word Completion for Text Documents*, n.d.). So, for example, if I used the word "language" enough times, LibreOffice's autocomplete algorithm would store that word in a collection. Then, as I type, the system tracks my keystrokes and when the pattern "lan" appears again, it will suggest an autocomplete of "language." Until more recently, and still probably in many word processors and text editors, autocomplete then relied on a rule-based, character-matching system, not machine-learning-generated statistical models. A particularly sophisticated autocomplete might also account for shuffling of letters, so it might recommend words that start with any permutation of the first three letters typed.

If we were to compare the two systems, we might say that autocomplete in word processors uses a list of words as a model and matches based on a series of rules or instructions that could be explained as follows:

1 Keep track of the frequency of words.
2 When the frequency of a word hits a particular numerical value, add that word to the autocomplete list.
3 Keep track of characters typed in each new word.
4 When the count of characters in a word being typed hits 3, search the autocomplete list for words that begin with the same first three characters and return them as suggestions.

No machine learning is used in such a system. In GenAI, the LLM does not actually contain any words—this is important as we will see in later chapters—but instead it contains weighted values that represent numerous features of language. It also contains some linear algebra. When GenAI systems construct text, they begin by encoding the tokens of the prompt into numerical values, it then passes those values through a series of linear algebra to predict the next most probable token, and eventually it is decoded back into a token and human-readable word. The system selects words and adjusts forward and backward across a sentence or even an entire document until the token limit is reached. Many GenAI systems will generate several candidate texts and reprocess them to determine which generated text fits the pattern best. The GenAI system is not assembling by letter, but rather by token following patterns of "meaning" that was extracted and encoded into the LLM as the model was being trained.

The distinction between searching a static list of words by letter and generating a series of words by matching meaningful patterns in language is important for two reasons. First, as I have emphasized, the GenAI systems contain LLMs that model the relationships between words in language and have been able to capture both syntax and meaning. We do not actually know what features of language have been extracted in the training of an LLM, but they seem to work and are able to generate meaningful text when used in an inference pipeline. Second, autocorrect holds a list of words and searches repeatedly through that list as the user types. The LLM does not contain any of the original text. It is a derivative of the original text produced by processing the original text data through a whole lot of linear algebra. The fact that these LLMs effectively model many characteristics of human language is the product of a considerable amount of trial and error. The training involves both automatic adjusting of numerical weights representing the tokens and a considerable amount of experimentation by researchers developing the models. For instance, they might experiment with the initial numerical values prior to running the training process or using different functions and regression techniques to adjust the weights as training occurs. Through this process of experimentation and stirring the data, the output eventually starts to look right. In short, LLMs represent human language in useful ways because of decades of work by researchers stirring a staggering pile of data and linear algebra "until it starts looking right" (Munroe, 2017).

Finally, while the LLMs do implement probability and statistics, the particulars of how they implement these are important. Predicting the next most probable token is not merely a matter of predicting the chances that one word appears next to another, but more importantly, it generates each token based on the mathematical formula of the LLM that represents a vast matrix of language features that capture the word, its meaning, and the context in which it appears. The interface between human language and the mathematical formula is occupied by an encoding system that converts language into

Figure 1.1 Machine learning ~ XKCD #1838.

numerical values. I will discuss this at length, but it is important that we do not simply handwave through the probability and statistics involved in GenAI, because the prediction is based on how language is modeled in the particular LLM that is being used, which, in turn, can tell us a great deal about the language itself. So, as we leave the MLA-CCCC's Task Force on Generative AI's working paper, it is important to note that the paper effectively describes the literature and makes good recommendations regarding teaching in a time of GenAI. However, the limitations regarding its discussion of the technology are critical and worth spending time addressing. I argue that with a more expansive discussion of LLMs, how they are made, and how they work within

NLG pipelines, we will be in a better position to discuss the risks and benefits of GenAI in writing and education and be able to position LLMs more effectively alongside our conversations in rhetorical theory. In the next chapter, we will turn to computing and the work done to build computer systems that can efficiently and accurately interact with natural language.

Note

1 I will discuss the notion of "intelligence" involved in generative artificial intelligence in the next chapter. While I am using the term GenAI since it is the conventional term, it is worth being skeptical of and even pushing back against the term "artificial intelligence" for the science fiction narratives that the term invokes. Imagining a hard, science fiction style AI like Data from *Star Trek: The Next Generation* or *Skynet* is inappropriate and inaccurate.

2 A Background in Computing

2.1 How to Think Like a Computer

In the previous section, I discussed the anxiety surrounding the risks associated with GenAI, particularly for educators using writing assessments in their institutions and classrooms. While some of the pedagogical responses are sound and useful, it is clear from the anxiety among educators and policymakers and from the MLA-CCCC's Task Force on Generative AI's working paper on the technology that there is an opportunity to expand our understanding of how GenAI works. However, I believe that we are in a better position to understand and discuss this technology in useful ways if we have a better understanding of technology, in general, and especially the discourses that produce it. The major challenge that we face when we engage with technology, as evidenced by much of the research in the field so far, is that if we do not deeply engage with the underlying systems, the engineering choices, and the discourse that constitutes that field, our understanding remains limited to the UI provided by developers. Since the UI was designed to provide an interface that makes technology usable and accessible, it often hides the inner workings of the technology itself. For instance, ChatGPT and other GenAI chat systems go to great lengths to make their interfaces seem like "chats" as we understand them. This makes them quite usable, but it significantly obfuscates the complex, integrated system that makes up the GenAI pipeline. I suspect that perhaps one of the most significant obstacles to effective scholarship in the Digital Humanities and Rhetoric of Science more generally is the foundational knowledge in technology and technical discourses. Since this seems to be such a significant limitation to the scholarship surrounding GenAI, I will begin this chapter by providing an overview of some of the key features of the fields of Computer Science (CS) and Information Technology (IT). This will reveal some characteristics of those discourses and will contextualize a history of how CS researchers, Data Scientists, and Computational Linguists have approached solving the problem of building a computing system that can interact meaningfully with natural language. This background will begin with a discussion of

DOI: 10.4324/9781003493563-2

computational thinking, then move to machine learning and artificial intelligence, and finally I will discuss the evolution of computer systems that can interact with natural language.

2.2 Computational Thinking

Since our goal is to better understand the academic and industrial fields from which GenAI arose, it is useful to focus on some of the more important threshold concepts—or concepts that are central to a discourse that give us a way to engage with that discourse (Adler-Kassner et al., 2016). Of these, none is more important to understanding CS than *computational thinking*. Computational thinking is an overarching framework behind the work that computer scientists do and how they approach problem solving. Jeannette M. Wing popularize the term in a piece published in 2006 in *Communication of the ACM*, a publication that represents the fields of CS and IT. This publication has a broad readership across academic organizations and the tech industry alike, and it represents the currents in the fields. In other words, if a concept like *computational thinking* is highlighted in this publication, it means that the concept is central to the fields associated with computing. In the piece, Wing (2006) defines "computational thinking" as a general concept that explains various activities that computer scientists perform as they develop technologies and that "involves solving problems, designing systems, and understanding human behavior." This means that work begins with problems that need to be solved. Once a problem has been identified, the computer scientist will design systems to solve the problem in a way that is informed by human behavior. Understanding technological progress as a problem-solving project is useful because we can gain deep insight into a particular technology and who it serves by looking at the problem it was created to solve, the people experiencing that problem, and the companies set to profit from the solution. Understanding technological progress as system design means that we should look at technology not as a single object, but as a system, and our understanding must expand across the entire system. So, we should not treat GenAI like ChatGPT as a single technology but rather as a complex system of interconnecting software and hardware. Finally, understanding technological progress as being sensitive to human behavior allows us to see how these complex systems exist within the matrix of humans, our patterns of action, our belief systems, and our institutions. With *computational thinking* computer scientists break down problems that humans face in their everyday lives through an iterative process of analysis, design, and development eventually yielding a system that will be successful if it works within some existing or new patterns of human behavior.

Since Wing foregrounded *computational thinking*, it has been a central focus of CS, especially CS pedagogy with its interest in preparing students for work in the field (Selby & Woollard, 2013; Shute et al., 2017). Selby and

Woollard (2013) provide an extensive review of the literature and explain that while the field has spent a great amount of energy discussing *computational thinking*, there is little consensus on what it is precisely. This is reasonable because it plays a significant role in defining the discourses of CS and IT on many levels, and therefore is not a single technique or strategy, but rather what Bourdieu (1977) describes as a disposition—which I suggest is a "*habit-of-occupying* a particular field in space and time" (Medina, 2018). In other words, *computational thinking* is the stance that CS and IT researchers and professionals take before they begin to interpret the situations they encounter. As a result, this disposition underpins the work that is done in the field, including work done to develop GenAI.

Within the literature on this disposition, three major themes appear: *computational thinking* is a "thought process" that involves "abstraction" and "decomposition" (Selby & Woollard, 2013). The notion that *computational thinking* is at its core a thought process suggests that it is work that involves rational effort to understand phenomena and solve problems, and it is worth highlighting the word "process." In Writing Studies, "process" should be quite familiar, and our own process movements saw writing as an iterative and recursive series of steps. The "thought process" in *computational thinking* is quite the same; solving a problem in CS involves an iterative process by which the solutions are gradually refined as researchers and engineers gain an increasingly profound understanding of the problem and as technologies advance providing more elegant or powerful tools to implement in the solution. For instance, the current state of GenAI is the result of incremental advances in mathematical modeling of language, advances in understanding of language itself, novel algorithms, and advances in hardware—particularly graphical processing units (GPUs) that are available to run machine learning operations. Since it is best to understand technological advancement as an iterative process, our understanding of LLMs and the GenAI systems built around them depends on some insight into the iterative process that has developed them.

Computational thinking uses two complementary heuristics that make it possible to bridge the space between the problems that humans experience that are being solved and the strengths and limitations of the technology that can be used to solve it. Those two heuristics are "decomposition" and "abstraction" (Selby & Woollard, 2013). Decomposition involves taking a large problem and breaking it into its constituent, solvable, parts. This might seem like a simple concept, but it is critical to computer systems because it is through this process that we can use technology to help us solve complex problems. Let us for a moment consider an NLP problem like finding the context in which a word appears in a text—or finding the concordance of a word. A human might do this by reading the text, finding every instance of the keyword, and recording it with the words around it, and the instructions one would need to provide to a human would simply be "find and record the 5 words before and after each occurrence of this word." The main limitations of the person's ability to accomplish this effectively are their attention (we make mistakes and miss things when we read) and the length of the text compared with their reading pace. These two limitations play precisely into computers' strengths. First, a computer can scan through massive amounts of text rapidly, and if the instructions written in code that the computer executes are correct, the computer will not make any mistakes in the task. In addition to the speed and accuracy benefits of using a computer for this task, finding concordances is not the most-exciting work, and if I can spend a little time writing a computer program to do it that can run while I go out and do something more interesting, I certainly will. The limitations of using the computer to do this task are mainly in writing the instructions correctly. These instructions must be considerably more specific than the ones I could give a person, and for that, the problem must be further decomposed.

If I were to decompose the problem, I might find three subproblems that each needs to be solved: first, finding the locations of the keyword in the text, second, finding the preceding and following words, and third, displaying the results in a useful way. The main obstacle within the first two subproblems is that computers do not process text by word, but character by character.[1] Once an approach is designed for each of these subproblems, it would then be written into code—or instructions that the computer follows to perform some operation. I will describe this in detail to illustrate decomposition further and because it will help us think about how computers interact with text which will serve us when we turn to LLMs.[2]

The first thing to consider is that computers interact with natural language initially as a stream of characters, so the sentence, "The quick brown fox jumped over the lazy dog," looks to a computer like a list of individual characters beginning with "T," "h," "e," and ending with "d," "o," "g," "." If I asked a human reader to pick out the word "fox," which I will call the "keyword," they would see the cluster of letters together and be able to pick it out immediately. Even if the words in the sentence were scrambled, we would likely be able to pick out the search string. If I wanted a computer to find the word "fox," one particular, brute force, approach would be to write instructions that define word boundaries, which would be space characters and punctuation, and then would iterate through the list of characters until it finds the character "f" that is preceded by a word boundary. Once that is done, it would step to comparing the second character of the keyword, "o," and if that is a match, it moves on to the third character of the keyword, "x," and finally it would try to match a word boundary. If this occurs, then the code would instruct the computer to do something in the case of a match. Otherwise, if at any point there is a character that does not match in the match sequence, the program would continue through the text and reset the character to match the beginning of the keyword. In the case of finding concordances, then, I would begin by writing instructions—or a program—that iterates through the input text and finds the location of all the matches of the keyword in the text. In other words, the first subproblem that can be decomposed out of the concordance problem is finding occurrences of a keyword in a search string and storing these locations so that we can use them later. The second problem is to search for the words preceding and following the search word. This gives us the context of the word. Once again, if I wanted the context to be three words to either side of the keyword, it would be easy for a human to see "The quick brown **fox** jumped over the" would be the window. The computing problem would involve continuing to iterate through characters using the word boundaries we defined before. In this case, the instructions for the computer would be to start at the start and end position of each occurrence of the keyword and iterate backward and forward until the configured number of word boundaries are found, so in this case three on either side. Once again, it might be convenient to store the starting and ending positions of each of these windows in which the keyword appears so that it is easier for the computer to find them again. Finally, we need to display the context in a useful way. The popular corpus linguistics too, AntConc (Lawrence, 2023), does this by displaying the sequence of words with the keyword and neighbor words in different colored fonts. A graphical user interface (GUI) like the one used by AntConc itself is a system solving a number of different subproblems including window-management, keyboard and mouse user interactions, font display, and so on. For a simpler command line interface (CLI) the output might simply display each instance of the keyword in its context on a new line with the keyword in all capital letters.

While the decomposition I've provided could certainly be implemented more efficiently and effectively, the notion that finding the context of all the occurrences of a keyword in a text involves more than one single process and when these processes are integrated, the computer can help the user do work that would otherwise take them an enormous amount of time. Consider finding all instances of "silly" in the script of *Monty Python and the Holy Grail*. Using the concordance algorithm that is part of NLTK (Bird & Loper, 2004), which, by default, finds the 79 characters around the keyword, this takes approximately 0.000983 seconds. The important idea here is that computer scientists use decomposition to create systems of instructions that allow them to use a computer to help them solve more complex problems. The main reason that we consider *decomposition* in our conversation around GenAI is to recognize that the process of developing GenAI has not been the product of writing a single app that could generate human-like text with LLMs, but rather it has been the product of solving numerous subproblems that taken together in a system can generate human-like text in response to natural language prompts. In other words, to describe ChatGPT as a single technology or single solution ignores the numerous subproblems that were solved, and when we consider the benefits and risks of this technology, we need to try to point to the exact feature of the technology and subproblem that was solved by it.

The second complementary heuristic within *computational thinking* is "abstraction" (Selby & Woollard, 2013, p. 2). Abstraction involves two processes and usually happens later in the development cycle once the problem has been decomposed and subproblems have workable solutions. First, "abstraction" involves generalizing processes or making a process suitable for solving similar subproblems within one project or across multiple projects. Second, it involves hiding layers of processing from other layers and users. Consider the decomposition above, one of the main functions iterates through a text and finds keyword matches. This same word search algorithm might also be used to highlight vocabulary words, generate word distribution and frequencies, and so on. Abstraction means writing the instructions in such a way that it can do so. If I were to abstract the functionality I describe above, then my program would need to take a text and a keyword as inputs and return a list of start and end indices for the locations of the word in the text. Andrew Hunt and David Thomas (1999) emphasize this notion of writing generalizable, abstract code because it prevents unnecessary repetition in the code making it neater and easier to maintain. In addition to making code easier to maintain, abstraction allows for what Bruno Latour (1999) describes as "blackboxing" (p. 183). Specifically, blackboxing refers to the way in which internal processes are hidden so that other members of an actant network need only know the process' inputs and outputs. For instance, when we use our smartphones to send text messages, we do not need to know the various processes involved in sending them. We only need to know the UI. Likewise, when computer scientists design algorithms, the users of those algorithms need only know what the expected input is to get a reliable output. If the word search algorithm is blackboxed, then the users of it only need to know what type of input to provide (text and a keyword), and they can count on a reliable output from the algorithm. At no point do they need to know how this works.

Abstraction and the resulting blackboxing are particularly relevant to the architecture of large, interconnected systems. When large systems are built—like those that are part of GenAI—they are composed of a series of functions each of which takes some input and returns some reliable output. This means that the system can be modular, meaning components like the LLM can be swapped out for a different one without having to rebuild the entire system. It also means that each component can be developed somewhat independently. This allows large communities of computer scientists, developers, and computational linguists to collaborate on projects. While "abstraction" is useful, it also raises serious ethical concerns (Durán & Jongsma, 2021; Latour, 2005; Martin, 2019; Quinn et al., 2022; Steen, 2015; Winner, 1993). Abstracting away processes and making it difficult or even impossible for the user to observe the processes that occur within it. Google search, for instance, is a simple webform that accepts a text input from a user and returns a list of websites with links and short descriptions. The process by which it does this is hidden and certainly includes things like storing tracking cookies in our browsers so we can be better targeted by advertisements. When we use some of the GenAI chatbots, we get responses to prompt, but within the system itself, this interaction might also be used to further train the system. It is hard for us to know for certain if the system itself is blackboxed. One of my goals for this chapter is to open the blackbox so that we can better understand how GenAI is working beyond the friendly UI.

The central disposition that CS pedagogy focuses on is *computational thinking*, and it was from this position that GenAI systems were developed. We will quickly move on to look at the systems and understand the problem that they set out to solve, the way that problem has been decomposed in a variety of different ways, and how systems of abstracted programs were assembled to create GenAI that is quite effective at producing human-like text. Before we go further, I need to briefly discuss Artificial Intelligence (AI) because there are numerous misconceptions about what it is, how it was built, and how it works.

2.3 Artificial Intelligence

While *computational thinking* is the disposition of computer scientists and other IT professionals, the product of their work is computer programs and computer systems. I described a program above that might find keywords in a larger text by following a set of instructions that have been encoded in some programming language. When the computer runs my program, it follows the instructions that define how it should iterate through the text, when it should store indexes of particular words, and so on. Regardless of how flexible a program might seem—just as GenAI seems to generate text dynamically based on the prompt it receives—ultimately, computers can only perform work following the instructions that have been written for them. Rules-based computer programs are programs where the instructions are written as a set of rules to follow. A keyword search program will follow a rule to iterate through a text, compare the character to the next character in the keyword, and store the index when the entire keyword has been found. These are fixed instructions, and as long as the input is of an appropriate format and the computer has enough resources to execute the instructions, it will do so without fail. Some of the systems that have been developed to deal with human language are rule-based, but the challenge with such systems is that every possible input must be accounted for in the rules. "Intelligent" systems are systems where instructions have been encoded, but some statistical model is used somewhere in the process. When we talk about "artificial intelligence" what we are talking about is systems that use some sort of statistical model within the instructions. Since GenAI is a system that uses "artificial intelligence," it is important to take a moment to clearly define it and how it differs from human intelligence. A clear distinction between the two allows us to avoid some of the risk of anthropomorphizing GenAI, and it gives us a better understanding of what occurs within the system when it generates text.

Intelligence is a challenging concept that has been implemented in highly problematic ways. When it comes to human intelligence, there is little consensus on what characterizes human intelligence, and when we attempt to do so the description of the phenomenon quickly loses its explanatory power and usually descends into bias (Pfeifer & Scheier, 2001). We might describe a student who learns course material quickly and recall it accurately as intelligent. We might also call a student who introduces new ideas based on interesting readings of evidence as intelligent. A student who is good at solving challenging problems would also be called intelligent. Individuals who are adept at understanding their own emotions and the emotions of those around them might be considered "emotionally intelligent" (Sternberg, 2000, p. 396). In short, a concrete and clear definition of intelligence in humans is difficult, but we generally do associate intelligence with human characteristics like agency, self-awareness, and free will. Numerous works of science fiction—too many to list here—play on the ways that artificial intelligence beings have or do not have these characteristics. This anthropomorphism that is interesting and entertaining in science fiction trickles into the scholarship around GenAI, particularly in those discussions that suggest we see systems like ChatGPT as a "collaborator" (Anderson, 2023). As I will discuss at length in the next chapter, thinking of GenAI as a collaborator is about as appropriate as considering weather prediction models collaborators. Consider, for instance, MetNet, a precipitation prediction neural network model developed at Google Research (Sønderby et al., 2020). This model uses similar machine learning processes to those used to develop GenAI, but would we consider this system a collaborator or even intelligent? Probably we would consider it a highly effective mathematical tool that is accurate at predicting weather based on a large model of past weather data. While GenAI tools like ChatGPT do seem intelligent in some ways, and it is probably in the interests of the companies that maintain them for it to seem this way, the AI of science fiction is not the reality now. The artificial intelligence available to us now in systems like ChatGPT is not "thinking" or "intelligent" in a human sense.

Pei Wang (2019), a leading researcher in the philosophy of intelligence in computer science provides a thorough discussion of how researchers in the field define intelligence in computers. He does this by outlining the difference between systems that simply do calculations—rule-based systems—and systems that show intelligence. The former require a complete knowledge set and sufficient computing resources to run the calculations in a reasonable time, whereas the latter have the ability to process information in variable contexts with insufficient knowledge and resources (Wang, 2019). This becomes clearer with a couple of examples. Some states—like the one in which I live—add sales tax on all items with some exemptions where the list of exempt items is complete. Local municipalities might also apply additional sales taxes for a range of reasons. Calculating the taxes and the total cost of an item represents a problem with a complete knowledge set. At the moment of sale, the price of the items, pre- and post-tax coupons, the various tax rates, and a list of exempt items are all known values, and so it would be relatively easy to build a program that could calculate the sales tax and total for an order and provide an itemized receipt. Since all of the information needed to run such a calculation is known, this is a complete knowledge set, and the program would not be an example of computer intelligence even if it does involve decision making (i.e. if an item is on the tax-exempt list, it's price would be excluded from the tax calculation). If any value in the knowledge set were to change, the system would merely need to be updated, and perhaps it could even request those values from some database that is kept up to date. In short, this is a rules-based system that is simply doing calculations because the knowledge set is complete, the number of variables is small, and a computer with sufficient processing power could easily do this calculation in a reasonable amount of time.

However, consider a problem like detecting people in a video feed captured by a home security camera to activate a recording for future review. To do this with a rule-based system, the program would need to have access to a knowledge set of every possible person-containing picture that it could match against. This, of course would be impossible. The environmental factors are variable and the shape and position of a human body that could appear in any given frame are so numerous that it would be impossible to create a complete knowledge set containing all environment/body/position combinations that could possibly exist. So, the first constraint of a rule-based system—having a complete knowledge set—is not satisfied. The second constraint of being able to run calculations in reasonable time would fail here as well. If we somehow had a complete knowledge set for this problem, it would take so much processing power to compare the current captured video frame with the set of images that this could not be done feasibly. Instead, this is a problem that could be solved by an intelligent system using probability. Identifying human forms in an image seems relatively easy to us because we are good at recognizing shapes within an image because we look at the image as a whole. A program would need to instead predict the probability of a particular pattern of pixels in the frame captured in the video feed contains a human form.[3] This kind of system is possible because it does exist and would be in the realm of what Wang (2019) describes as intelligent because it can make calculations based on incomplete knowledge in real time. Home video surveillance apps can detect a human shape in the frame and can do so quickly enough that they activate recording when they do.

I will refer to the former system that relies on a complete knowledge set as a rule-based system and systems that can work with incomplete knowledge set as machine learning (ML) systems. Rule-based systems and ML systems have both been used to do computations with human language, and they are both part of GenAI systems as we will investigate later in this chapter. In the field of ML, approaches generally fall into two camps: supervised learning and unsupervised learning. Supervised learning means that data is labeled in some way by a human researcher prior to training. For instance, early customer service bots used intent recognition training. Intent recognition involves gathering a dataset of relevant utterances and labeling those utterances based on some predefined categories. A company might building a customer service bot might begin by recording customer calls and labeling questions based on the topic or intent of the speaker. All of the calls asking for help with a password might be labeled "password reset request." Once a dataset is gathered and labeled, it is split into training and evaluation groups. The training group is run through a neural network, which is a lot of linear algebra, which gradually adjusts until it becomes a model that mathematically represents the categories in the dataset. The evaluation group is then used in reverse fashion to confirm and further adjust the model. When the model is deployed, the user input is processed through the neural network model and most similar category is identified. A user might ask for a password reset in a lot of different ways, but the model will be able to approximately identify that all of these ways fit into the password reset category. From there, a rule-based response might send the caller an email with a password reset link and the system will play some pre-recorded message informing them to check their email. With supervised learning, someone needs to spend hours labeling language data. One of the considerable ethical concerns of AI that is less frequently discussed is the salaries paid to the workers who have labeled natural language data. This work "is often outsourced to labour markets in the Global South where companies can find workers who are fluent in English and willing to work for low wages" (Taylor, 2023). I will discuss ethical concerns in the fourth chapter.

Unsupervised learning is ML in which data is cleaned and input into a learning algorithm and a model is built on naturally occurring patterns within the data. There are a few prerequisites for this to work: the data must be represented mathematically because the model itself is a lot of numerical weights and linear algebra. Second, there must be enough data for reliable patterns to emerge. Finally, we must have enough processing power and time to run the learning algorithm on the massive data pool. The first step in unsupervised ML is cleaning the data and converting it into numerical values if necessary. Then a mathematical model is selected, and the coefficients of the variables are gradually adjusted by applying and comparing input and output until the model fits the data. Linear regression is often used in machine learning and can be used to automatically model phenomena that have linear relationships, when means there is a relationship between an independent variable and a dependent variable. Modeling natural language is significantly more complex, but a linear example will be illustrative to the general technique. Let us assume for a moment that we have an exam where student grades depend only on how many hours a student studies. The hours studied is the independent variable and the grade is the dependent variable. We administer this exam 100,000 times and we want to figure out a model that will predict the grade of a student based on the number of hours they study. This model could be a suitable candidate for a standard linear function

$$y = \beta_0 + \beta_1 x$$

We will assume that x stands for the number of hours studied and y is the exam grade. Since it is unlikely that every single point in the dataset will fall perfectly on a single line, the training algorithm will do linear regression by gradually adjusting the values of β_0 and β_1 until the sum of the distance from the line drawn by the function of every point in the dataset is minimized. After linear regression, the result is a function with values for β_0 and β_1 so that when a particular number of hours of study is assigned to x we can do simple arithmetic to calculate the grade that student should get based on the prediction of our model. If the model is a good one, the prediction will be accurate. The "learning" component of ML occurs as the function is gradually adjusted based on the dataset. This particular function has two dimensions (x, y) and can be drawn in two-dimensional space, but student grades on exams are probably not particularly well modeled by just hours studied. Instead, there are numerous other factors that contribute to the grade on an exam, and so the neural networks are used to model the relationships so that instead of having two parameters that are adjusted in learning, there might be many. LLMs, for instance, might have billions of parameters. OpenAI's GPT-3.5, one of the large language models available in ChatGPT, has 20 billion parameters.

There is considerably more math that one could discuss in ML, and if one were interested in diving into this topic more profoundly, they would be well served by Andrew Ng's courses which can be found readily online. Likewise, Stephen Wolfram (2023) offers an excellent overview of ChatGPT and its math. The important thing to focus on before we shift more explicitly to language is the notion of rule-based programs and intelligent programs. Rule-based programs work because the developer can write rules into the instructions for the computer that account for all of the potential cases that could be handled. They rely on a complete knowledge set that can be used in real time. Intelligent systems are required for tasks like predicting grades or identifying patterns of human forms in video feeds or generating natural language because they all deal with incomplete knowledge sets. It is impossible to model every possible grade outcome because there are so many variables that determine them, there are unlimited variations of what a human form in a video feed could look like, and the combinations available in language that produce meaning are both numerous and ever changing as language evolves. An intelligent program is not a program that "thinks" in any sort of way. When a person uses language that language could be modeled, but the person accounts for the language, the extra-linguistic context, the particular relationship they have with their interlocuter, what occurred earlier in the day, and their desired outcome for this interaction. GenAI, however, merely has an exceptionally large model of language that has been trained on a truly staggering dataset and that is baked into a larger system. GenAI is a phenomenal technology, and it is an example of an intelligent system, but it is important to note that it is not intelligent in any human way—at least not yet.

Notes

1 Computers can process text word-by-word if that word has been encoded into some numerical value as I discuss below.

2 In the discussion, I intentionally present an unoptimized approach to solving this particular problem because I want to illustrate the way that computers interact with language.

3 ReCAPTCHA challenges that ask users to prove they are not a bot by identifying bicycles or stoplights in images work because of the difference in how computers and humans process images.

3 Computers and Language

3.1 Rules-Based Approaches

I spent the time in the previous chapter discussing *computational thinking* and AI so that we can begin to understand how computers might be programmed to solve a problem like running computations on an incomplete knowledge set like natural human language. This "language problem" exists because while language is conventional and regular in many ways, it is also large, diverse, and dynamic. The permutations of language use are enormous and difficult to model because words have flexible meanings that change over time and might have multiple meanings depending on the context and their relationship with other words. In addition to this, language includes devices like irony and discourse-specific usages that might not find their way beyond a small circle of language users and might be very different from their conventional usages. All of this makes handling human language an exceedingly difficult problem for computing. Computer Scientists and Computational Linguists have addressed this challenge with both rule-based systems and ML systems—both of which are involved in modern NLP and NLG systems. There are two major approaches by which the language problem can be solved: (1) constrain language to a sufficient extent that it can be contained within a rules-based system or (2) use ML techniques to construct a computational model of language.

Rule-based solutions to the language problem are interesting for a few reasons: first, they do not require extensive model training or expensive processing power for running computations. Examples of a rule-based systems are simple word search programs that find instances of words in a text or list-based autocomplete features like I described in the first chapter. Second, the output of a rule-based system will be absolutely reliable. This is important to note: in rule-based systems, given a particular, correct input the output will be predictable. For instance, a rule-based system that calculates sales tax of 8% would always return 8 if the input is 100. To implement such a system the developer must define explicit instructions or rules based on every possible input. To use a rule-based system to address the language problem, the language must be

DOI: 10.4324/9781003493563-3

constrained so that a rule can be written for every possible construction. A phone tree that you might encounter when trying to call customer service is an example of such a system. These systems play a prerecorded voice message with a number of options. The user selects from one of the options and the system continues to either direct the call to a human agent or respond with another corresponding prerecorded message. In this case, the choices are human language we can understand, but the utterances are constrained to the options provided by the people who designed and built the system. Likewise, role-playing video games in which the user's character interacts with non-player characters by selecting utterances for the character to say follow a similar rules-based model where the language choices are constrained to a limited, predetermined set of options, and the non-player characters respond based on the rules that correspond to the choice. In the first example, such systems are particularly frustrating, I suspect, because they are *not* language when we want it. We want to speak with a human, and we get a program.The second example is not exactly language use in a conventional sense insofar as interlocuters have a considerable feelings of agency over what they say, but if the player is immersed in the game sufficiently that the choices offered by the game resonate with the player, they might feel like they are having a conversation within the game. While these are quite distant from LLMs, I do suggest that they are worth considering because they do allow the users to interact with the computer with something that approximates human language. Moreover, in cases where the language can be limited in this way, they will execute much more efficiently and quickly than intelligent systems.

These initial examples are systems that only barely allow computers to process human language. The computer actually is not processing language, and instead, it is working with a number that corresponds with the choice that is written in human language for our benefit. A more interesting example of a rule-based system that processes natural language is the one that appears in the 1977 text-based adventure game *Zork* (*Zork Source Code*, 2020/1977). *Zork* is a game of interactive fiction in which the player is displayed text describing a scene and action that are part of an ongoing narrative. After each description, the player is prompted to input text through a plain text interface. Based on the input text, the game guides the player through the story. Unlike the phone trees or role-playing games described above, *Zork* relies on the player inputting natural language which is interpreted—or at least parsed—by the game as it advances. The game works because all of the possible permutations of *valid* user input are defined within a series of decision trees. In other words, users can input natural human language and the game advances based on rules that match up outcomes with inputs. Since human language is vast, *Zork* does not attempt to represent all of it in the rules, but instead it constrains language into a set of viable commands that follow the structure Verb, Direct Object, and Indirect Object. Valid user input follows this structure and begins with valid combinations of verbs, objects, and direct objects. If the input is invalid, the system alerts the user to try again. The programming language used to handle this work is called the Zork Implementation Language (ZIL). In *Zork*, the work of processing the input language is started by the parser, which is the technology that is involved in solving the language problem. Steven Eric Meretzky explains in the *Learning ZIL* manual, "There's a notorious part of every [interactive fiction] program called the parser. It gets the first look at the input. If it decides that the input is indecipherable, for any of several reasons, it handles the input" (Meretzky, 1995). This means that the parser is responsible for interpreting the natural language. The parser begins with the first word and matches that with a list of possible verbs. Then it moves on to the object and so on. If, at any point, the parser does not find a matching rule for part of the input, it returns an error message to the user. Once the parser is done, either an error message was returned or the parsed verb, direct object, and indirect object are passed further into the program for matching. The rest of the program matches predefined patterns of verb and object with predefined messages that navigate the player further through the narrative paths.

What is most interesting about this program is that the player can create their own complete commands in human language and not simply choose from preexisting options. From this human language input, the program returns reliable responses based on the rules defining how to handle particular patterns of input. The program contains all of the permutations of possible paths that the player could take through the game, and the program using the tightly constrained model of language and ZIL to interpret the player's input and advance them through the game or raise an error message if the input does not yield a close enough match. Early chatbots like ELIZA which matched keywords and returned responses rooted in Rogerian psychotherapy and A.L.I.C.E. which used a heuristic pattern matching algorithm to return preprogrammed responses both use similar approaches to *Zork* (Wallace, 1995; Weizenbaum, 1966). Constraining language around a predictable set of rules and keywords yielded useful but limited results.

Another rules-based approach that I want to discuss relies on syntactic patterns that occur within human language. Systems like this model language are based on parts of speech and syntax. Seb Pearce's *New Age Bullshit Generator* (n.d.) is an example of this. If the user clicks the "Reionize Electrons" button at the top, one receives a series of paragraphs that bear a striking resemblance to new age discourse. Here is a sample from one iteration of the site: "The infinite is buzzing with frequencies. Today, science tells us that the essence of nature is rebirth." Seb Pearce explains the site in a post added in December 2022, "The reason I made it was to demonstrate how easy it is to fool our brains with this fact-free, emotionally charged language that tries to bypass our rational thinking." Pearce has released the source code for the site, and it is available on GitHub. The *New Age Bullshit Generator* produces language by randomly selecting from sentence patterns like, "nMass is the driver of nMass" and "We exist as fixedNP." The placeholders "nMass" and "fixedNP" and others are replaced by randomly selected items from the corresponding list in a collection that is appropriately called "bullshitWords." The system randomly selects from the various sentence structures in a predetermined pattern and replaces the placeholders with randomly selected "bullshitWords" and then has some post-processing to cleanup extraneous punctuations and spaces (*Sebpearce/Bullshit: Generates New Age Nonsense on the Fly*, n.d.). Once again, the entire range of possibilities within this model of language is predetermined in the code.

We could certainly argue that these rule-based approaches are not actually dealing with "natural language" because all possible inputs and outputs are predefined, or they randomly construct language that is only meaningful insofar as we can interpret them. However, these approaches reveal a great deal about some of the ways that computers can interact with human language. Computers require rules to perform operations and at this point cannot infer or intuit appropriate responses to data for which a rule was not written. Language, however, is exceedingly difficult to model with rules because of the size, changeability, and inconsistencies it contains. Rule-based approaches need to constrain the language to fit into some discrete set of rules. In the case of *Zork* and early chatbots, these rules have to do first with parsing the input language into sets of verbs and objects or keywords and matching an appropriate, corresponding response. In template-based language generation, language is constrained into sets of syntax and vocabularies. These are randomly assembled mad-lib style generating text that can only ever follow the patterns encoded in the templates. Rules-based systems are certainly less advanced than intelligent systems, but there are advantages to template-based systems, in particular. Such systems can be highly effective and perform as well as ML-based NLG (Deemter et al., 2005; Reiter, 1996). Template-based systems function particularly well in rhetorical situations where the genre is particularly stable. These systems produce reliable output and are not subject to the sorts of hallucinations that appear in GenAI systems. Further, rules-based systems do not require vast computing resources. According to an analysis reported by *Business Insider*, ChatGPT in 2023 costs more than $700k *per day* to run (Mok, 2023). Most of this cost is in computing, which is done in some cloud infrastructure where customers can pay a company like Amazon or Google or Microsoft to run programs on their hardware infrastructure. At the time of writing, the smallest server instance on AWS, Amazon's cloud, designed to perform NLG tasks costs $3.06 *per hour* (*Amazon EC2 P3 – Ideal for Machine Learning and HPC – AWS*, n.d.). Rule-based programs require much less expensive infrastructure because they only require pattern matching and simple logic and not processor intensive matrix multiplication that is part of GenAI.

Rule-based systems, while significantly different from ML systems, illustrate a foundational approach to solving the language problem. They position language as a deterministic, constrained structure. This structure is based on matching patterns either in vocabulary or grammar, and they are successful in contexts where a significantly constrained subset of language is sufficient. Before moving on to "intelligent" systems, it is worth considering rules-based systems in terms of meaning making. In these systems, computers have little to do with the making of meaning. The constraints placed on the language are determined by the developers, and even in systems that use randomization to assemble language, the constraints of the randomness is predetermined by the developers. For the computer, it is clear that the words have no meaning. I emphasize this here because the main limitation of rules-based systems is that the available meanings that can be produced must be predetermined and therefore cannot achieve the variety and variability of meaning that occurs in natural language. Likewise, the choices in how language is constructed are prescribed in a sequence of rules. There is no conscious agent to be aware of any rhetorical situation. In the next section, we will see that ML approaches to NLG come closer to natural language and even encode some meaning in the models that define how language is generated.

3.2 Encoding Language

Rules-based systems have hopefully illustrated the nature of the language problem: they cannot effectively work with textual data in the same way that we can. While computers may be able to far exceed us in their ability to run calculations, for them to do so, the data must be converted into numerical values. Rules-based systems get around this by matching patterns of letters. When I tell *Zork* to "Go East," it is matching the characters of the words with ones that have been saved as keywords in the lists. The conversation of text data to numbers is done character by character in this case, and there are a number of character encoding systems that might be used to do this including *Ascii* and *UTF-8*. These encodings treat characters as numerical values, and none of the meaning of the word is captured in the encoding. However, if computers are to operate on the meaning of these characters, something that does not readily translate into numerical values, then some other system needed to be developed to encode that meaning into a numerical format. Encodings like the ones used in intelligent NLP and NLG systems solve the language problem by attempting to convert or capture features of language in numerical values.

The simplest way to begin moving beyond encoding characters to encoding semantic units like words is to assign an arbitrary numerical value to each word. Imagine that a particular language only needs to contain four words: "apple," "horse," "eats," and "hay." An arbitrary encoding system might assign values where apple = 1, horse = 2, eats = 3, and hay = 4. The sentence "The horse eats hay and apples" could be encoded "2341" if we leave out the words "The" and "and" as "stop words" or words that are quite common but carry less semantic weight. The first problem with this is that when calculations are made on these numbers, larger numbers carry more weight, meaning, "hay" would carry the most weight regardless of how important it is to the language or the document being analyzed. One option would be to weight words differently based on their importance—and that is actually what happens with some ML applications—but the issue of assigning importance is difficult. The second problem associated with encoding is that words do not have single meanings, and human readers understand meaning of words not individually but in the context of the sentence, paragraph, entire document, and even the social context in which the document was produced and received. This is obvious in the two sentences, "I took my dogs for a walk" and "I went for a run, and now my dogs are barking." The word "dogs" appears in both sentences, but to the human reader familiar with the particular way I am using English, they have different meaning (an animal and my feed respectively). If some numerical value were to be assigned to "dogs" then to the computer, both uses of the word would be encoded to the same value. Clearly, a simple arbitrary encoding system does not account for these two issues and has other issues as well. Any predictions made on such a model would be ineffective.

Since assigning numerical values arbitrarily is not useful because it results in calculations that are weighted in ways that do not capture the meaning of the words, it is also possible to assign numerical values to words based on how important the word is to the text itself. By encoding the importance of the word, this encoding system allows the resulting values to capture more information about the words. They now contain not only the word but also the importance of the word or the value of the word in the document. From that information, a system could predict what a text might be about. One highly popular way to assign value to words in a document is the Term Frequency Inverse Document Frequency (TF-IDF) algorithm introduced by Karen Sparck Jones (1972). The TF-IDF algorithm has commonly been part of search engines, text summary systems, and word clouds, all of which gain insight into text by identifying the most important words in the document. The purpose of this algorithm is to define the significance of each term in a given document in relationship to its prevalence in the documents in a corpus—or a large, representative collection of usually written, natural human language. There are a number of different variations to the TF-IDF algorithm, but the key to them all is that they are interested in word distributions and frequencies. One of the term frequency calculations that is easiest to understand is the raw count of the term divided by the total number of terms in the document. Term frequency within a single document is not enough to tell us how important a word is to the document because some words appear frequently in all texts. When the term frequency calculation is combined with the inverse document frequency calculation the identity of words that are particularly important to this document can be seen. Inverse document frequency is calculated by dividing the number of documents in the corpus and dividing it by the number of documents containing the term. Usually, the inverse document frequency calculation is then passed through the base 2 logarithm, which has the effect of decreasing the weight of the IDF term. An example of this calculation will likely make it somewhat clearer: imagine the TF-IDF was being calculated for the word "genre" that appears 25 times in a 500-word document that is part of a larger, 10,000-document corpus. In that corpus, let's imagine that "genre" appears in ten documents. The calculation would be $(25/500)\times\log_2(10{,}000/10)$, which yields approximately 0.498. If we want to use the TF-IDF algorithm for something like topic modeling or summarizing or search results, we would find the score of each word in each document. For topic modeling and summarizing, the highest scoring words would probably tell us something about the topic or what the document is about. For search results, we would rank search results by how many of the search terms appear as the most important terms in the document. So, if I search for "rhetorical genre theory," the search engine would return documents or web pages where the words "rhetoric," "genre," and "theory" have the highest TF-IDF scores.

Since the initial subproblem within the language problem is how do we encode language into numerical values, TF-IDF illustrates one way that can be done. Words now are represented not as a series of characters, nor by arbitrarily assigned numerical values, but as values associated with some meaningful information. This algorithm is useful for a number of NLP processes including identifying the topic of a document. While the topic of a document is more than simply the words, but rather is the meaning of the words together, identifying the most important words will likely provide a pretty good guess as to what the topic of the document is. Likewise, the algorithm can be used to help calculate lexical density. A document containing a higher percentage of high TF-IDF values will have a higher percentage of significantly meaningful words, meaning it will be more lexically dense. This illustrates the ways that this single calculation can be used to make other inferences about the text. For instance, TF-IDF helps us calculate lexical density, which, in turn, can help us place a text in different domains—academic writing tends to be lexically dense, whereas popular writing tends to be less so. This algorithm has been so effective at identifying the topics of documents that in 2015, it was part of 83% of library recommendation systems (Beel et al., 2016). Despite its uses, TF-IDF and many NLP techniques focus on words as atomic units and struggle to capture the meaningful relationships between words. It is also worth noting that ML techniques are not used in TF-IDF calculations, and so technically this step into encoding language is rules-based.

While TF-IDF illustrates how additional information from language might be encoded, it does not effectively capture the meaning of language, particularly contextual meaning. It also does not help model the complex relationship between words, and so it does not help very much with NLG. This approach to encoding language treats language as a collection of individual words and it is good for modeling the main topic insofar as the most frequent specialized words within a text will tell us what a text is about, but it will not tell us the meaning of the text. For instance, one could imagine two essays on the same topic from vastly different perspectives and following different lines of argument, but analysis with TF-IDF would yield the same topic for both texts. Since the meaning of language depends on the situated relationship between words, for language to be encoded effectively, it needs to capture that relationship. This is done with word embeddings, and GenAI systems rely heavily on them.

3.3 Word Embeddings

Unlike frequency calculations that encode words into single values, embeddings convert them into a number of values determined through a process of training using ML techniques. Since a text can be split a number of ways (by word part, word, sentence, etc.) there are a number of types of embeddings, but they follow similar approaches. Word embeddings are vectors that represent the position of the word in vector space. Vector space is an n-dimensional space containing vectors—directed lines from the origin to a particular point identified by n-coordinate values. The simplest way to think of this is in three-dimensional space. Consider the following words: "student," "god," "eagle," and "witness." I have selected these at random. *Gensim* (Řehůřek & Sojka, 2010), a popular framework for topic modeling, provides a Word2Vec training algorithm and the Lee Background Corpus that is included with the framework, three-dimensional word embeddings were trained that yield the values below (Řehůřek & Sojka, 2010). The training was done on unlabeled text with the algorithm identifying hidden features which I have labeled "Feature 1," "Features 2," and "Feature 3." What these features are exactly is hidden from view, and I will discuss how these features were identified below in the section on Word2Vec and GloVe. For now, we might think of them as particular characteristics of the word that have been extracted.

To visualize the relationship between the words, these values can be plotted into a three-dimensional vector space as vectors—or directed lines beginning at coordinates (0, 0, 0) and ending at the coordinate value for the particular word. Figure 3.1 shows these vectors plotted on a graph. The relationship is

Table 3.1 Word Embeddings Graphed in Three-Dimensional Space

	Feature 1	*Feature 2*	*Feature 3*
student	−0.20483153	0.21346898	0.13723841
god	−0.2763081	0.20785713	−0.10340062
eagle	−0.1720139	0.33277208	0.2535168
witness	−0.21861959	−0.17671889	0.4436871

determined by how close the words are together with word vectors of more closely related words being closer together, which results in a smaller angle between the two vectors. The particular way these values are generated by extracting features from the training data will be the focus of the next several sections. When word embeddings represent language, it means that the vectors are build where the values they contain each represent some feature of the word, its meaning, and its relationship to the other words in the training data. Likewise, when words are converted to vectors, which are numerical values, computers can process them, and as vectors become larger capturing more features of language, the computer gains access to more information or more of the meaning of the language. In the example I have provided above, the vectors only contain three values, meaning not very much information has been extracted. I chose to use three dimensions so that it could be plotted on a graph more easily. However, since we are not necessarily limited to three dimensions, word embeddings may have many more than that. The greater the dimensions, the more features can be extracted from the language. For example, if I were to label features of human words manually, I might choose age, education level, and notoriety as features giving the word "child" low values for all three. The word "adult" will probably have higher values for the first two since adults are older, and generally we think of them as having higher levels of education, but probably not much notoriety. The more values in the vector the more space there is to encode characteristics for each word, meaning more information can be encoded in the vector. At the same time, the higher dimensionality of the vector the more processing power is required to perform operations. While I describe manually selecting features to encode as an example to illustrate how various features of a word might be captured in a vector, the main difference for ML word embeddings is that the features are extracted automatically and we don't necessarily know what information is contained in each feature, and these embeddings might be of very high dimensionality. GloVe, an approach for constructing word embeddings that I discuss below, provides a set of pre-trained word embeddings with a dimensionality of 300, meaning they are vectors containing 300 values (*Stanfordnlp/GloVe: Software in C and Data Files for the Popular GloVe Model for Distributed Word Representations, a.k.a. Word Vectors or Embeddings*, 2015).

The calculations done in GenAI are performed on word embeddings, so we can look more closely on how meaning has been encoded into these vectors of values. This encoding is not done manually. Instead, it is done automatically by various training algorithms. The easiest way to do build vectors is using the "one-hot" encoding technique, and this technique is often the initial state used by many ML training algorithms. This begins by establishing a matrix—which is an *n*-dimensional grid—where each cell contains a value, each row represents each word in the vocabulary of the corpus and each column represents every other word in the corpus. For each row, the value 1 is assigned to the cell where the word and the column are the same word and

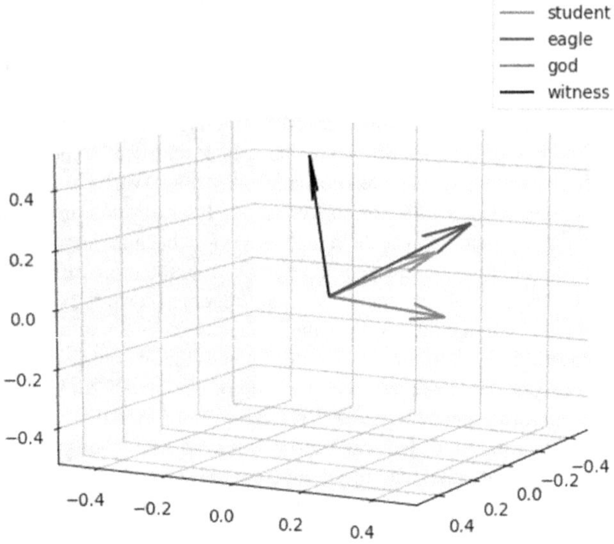

Figure 3.1 Words represented in three-dimensional embeddings.

Table 3.2 One-Hot Encodings Sometimes Used as Initial States for Machine Learning

	Apple	*Horse*	*Eats*	*Hay*
Apple	1	0	0	0
Horse	0	1	0	0
Eats	0	0	1	0
Hay	0	0	0	1

zeros for all other cells. Returning to the four-word-language that I described above, the one-hot encoding look like this:

In this encoding, "apple" would be represented as (1, 0, 0, 0). This effectively solves the problem with inappropriately heavy weights of some words and light weights of other words that occurs when we arbitrarily assign numerical values to each word. It also solves the problem of not being able to capture much information when a single number represents the encoding of the word. The challenges with one-hot encoding are that the table will be very immense, and the encoding does little to capture the meaning associated with the words or the relationship between them. However, it does illustrate a way we might encode the vocabulary into a matrix where each row represents a word vector, and each has a unique value without arbitrarily heavy or light weights for any individual word. This encoding system does not use ML and no features are captured except the existence of a word in the vocabulary. It does provide a starting point for better word embeddings that are produced by ML. The techniques I will turn to now use ML techniques to iteratively modulate the values of each value in each row of the matrix in an attempt to automatically capture more meaning about the word and the relationship with the words around it. Once words can be encoded, interesting NLP and NLG processes and solutions to the language problem can emerge.

3.4 Word2Vec & GloVe

With one-hot encoding, numbers were assigned in a binary manner (every value is a 1 or 0) and only capture the presence of a word, but this could be used with the TF-IDF algorithm so that instead of assigning 1 or 0, the TF-IDF value for each word in a document would be used. However, this would only yield word embeddings that are valid for a single document and would not create a vocabulary model that is valid for the entire corpus. In the early 2010s, researchers made a number of significant breakthroughs in machine learning algorithms and in the hardware that allowed them to develop better approaches to generating word embeddings in a way that better captures the meaning of each word in context. The Word2Vec (Mikolov, Chen, et al., 2013) and GloVe (Pennington et al., 2014) algorithms both rely on the fact that meaning of human language is connected to the relationship between words and their proximity with other words. It is also important to note that researchers used trial and error and experiments to develop these, which means the models are based on patterns that occur naturally in language rather than imposed by the researcher, and these patterns are detected using unsupervised training rather than having a researcher assigning values manually. The ML involved in these approaches requires a large set of training data because the data must me representative of language. How this corpus is constructed and what part of "language" does it represent will be topics that I return to in both the third and fourth chapters. Once a corpus has been collected, the first step in both of these approaches is preprocessing. While word embedding is a subproblem within the language problem, generating word embeddings is further decomposed into even smaller problems. During preprocessing, extraneous data including things like HTML tags if the data was scraped from the internet or page numbers or other text that is not really central to the text itself are all removed. The remaining text is tokenized and perhaps stemmed. Tokenizing a text is the process by which the larger text is broken up into semantic units. It is important to note again that a token *is a semantic unit*—a word or word part or punctuation mark that has meaning. This is significant because the units that NLP and NLG are operating on are tokens and not characters which means these algorithms and processes work with semantic units and not phonetic units.[1] This process of tokenizing also explains how GenAI predicts the next most probable unit of meaning. Tokenizing texts can be done in a number of ways and can be as simple as splitting a text using punctuation and spaces as token boundaries or it can be complex using regular expression matching—or character pattern matching—or even statistical analysis. Some tokenizers will split a text into prefixes, suffixes, and roots like splitting "antihero" into "anti" and "hero" as two distinct tokens. Likewise, there are sentence tokenizers and word tokenizers that split the text into sentences or words. The particular tokenizer that is chosen is important because how a text is broken into discrete units has significant implications for the way that language is modeled, and like any of these topics, further research and analysis into the rhetorical properties and pedagogical implications of each tokenizer would be worthwhile. In addition to tokenizing, preprocessing might also involve the morphological process of stemming. Stemming converts various tokens into the single root. This would mean the words "talk," "talking," "talks," "talked," and "talker" all into the stem "talk." Once again, there are numerous approaches to stemming and it is significant because it allows multiple forms of the same word to be represented by the same token. This is significant if we consider for example the two conjugations of the verb "to talk" in "I talk" and "one talks." We see that the two conjugations carry the same meaning but are conjugated to different subjects, but to a computer "talk" and "talks" do not equal each other and without stemming they would be treated as separate words.

Once the corpus has been preprocessed, the tokens can be introduced into an ML system that automatically generates word embeddings for each token. In 2013, Google researcher Tomas Mikolov and research team developed an approach to vectorizing that captured the context of the word as well as the word itself. This approach called Word2Vec (Mikolov, Chen, et al., 2013; Mikolov, Yih, & Zweig, 2013), depends on two particular techniques: Continuous-Bag-of-Words (CBOW) and Skip-Grams. The first step involves building a vocabulary from the preprocessed data. A vocabulary is a set of unique words that are contained within the corpus meaning. If the preprocessing involved stemming, then "talk" but not "talks" or "talked" will appear in the vocabulary. For each word in the vocabulary, a random word vector of a certain dimension will be initialized to represent the starting state of the word. Then the algorithm iterates through the corpus looking at spans of words. Each span is a window that contains a target word and the several words on either side. The random vector for the target word is then iteratively updated by predicting both the target word vector based on known vectors of the surrounding words and predicting the surrounding word vectors based on the target vector. The vectors are all updated by comparing predicted word vector with the corresponding actual word vector. With enough iterations of training—known as epochs—the word vectors gradually adjust so that the predictions become increasingly accurate, and each vector captures the semantic and syntactic meaning of each word as its relationship to other words in the window. Once this training process is complete, the result is a

Word2Vec model that can provide word embeddings for each word. It is worth noting that the model training only needs to happen once, although it can be updated or even retrained, and the word embeddings are available for any other number of NLP or NLG tasks.

Mikolov Chen et al. (2013) discovered some interested patterns emerged in the resulting vectors. One such example is the ability for the models to capture analogous relationships. For instance, they discovered that the following expression is true:

$$\text{vector}(\text{'King'}) - \text{vector}(\text{'Man'}) + \text{vector}(\text{'Woman'}) \approx \text{vector}(\text{'Queen'})$$

This could be read as if we take the vector representation of the word "king" and subtract the vector representation of the word "man" and add the vector representation of the word "woman," we get a vector that is remarkably close to the vector representation of the word "queen" (Mikolov, Chen, et al., 2013, p. 2). In this particular word embedding model, the difference between "queen" and "king" is a difference in gender, and "king" is to "man" as "queen" is to "woman." While this may not seem like a significant discovery to a human reader because our understanding of language is built around these kinds of relationships, what is particularly impressive about these word embeddings is that additional meaning about the relationship between words has been encoded in the vectors. Word2Vec is quite effective in extracting information from text, but it suffers from a few limitations. Since it calculates the word embeddings based on a relatively small window, it assumes that the meaning of words is largely in the relationship they have with the words that appear immediately around them within the phrase or sentence. Often this is not the case and that the encoding of the word should depend on the relationship it has within the larger context of perhaps the entire document.

Shortly after the introduction of Word2Vec word embeddings, Jeffrey Pennington et al. (2014), introduced GloVe or "Global Vectors" model that allowed information to be extracted from the entire document. The approach to calculating the word vectors is quite different, and instead of generating them based on calculations and predictions of individual words and the words immediately around them, the vectors are constructed by calculations based on the entire global context of the document. The GloVe training algorithm's goal is to improve the performance on analogy tests, or the ability of the model to identify and complete appropriate analogies (Pennington et al., 2014, p. 1533). This model is built by first initializing a matrix of values that count the co-occurrences of words across the document. In this matrix, the rows represent each word, and each column is a word that appears somewhere within a predetermined window centered on the target word. The algorithm then calculates the probability that each word will appear with any other word in the corpus. For instance, if the algorithm is calculating the vector for the word "write," then it would more likely occur with the word "paper" than it would with some other word like "fashion." The algorithm then calculates for each target word the probability that a particular co-occurring word will occur in any window surrounding the target word throughout the corpus. The information that is extracted then focuses on the global relationships with between words. The resulting word embeddings follow the pattern where more closely related words will appear closer together when graphed in vector space, and information about these relationships can be generated by calculating the angle between the vectors where the smaller the angle the more closely related the words.

Word2Vec and GloVe both approach encoding language in very different ways, but they share in common the ability to convert words into vectors that contain a number of values each representing some characteristic of the word and its place in language. Word2Vec encodes words based on the relationships they have with neighbors in a local window and assumes that information about word depends on the immediate context of the word. GloVe encodes words based on the probability of co-occurrence—or the likelihood that words will appear next to each other—across the entire corpus. Word2Vec uses neural networks and ML techniques to adjust initialized values in word vectors whereas GloVe uses statistical analysis and matrix factorization. In both cases, these algorithms follow a different approach to the language problem than rules-based systems did. Rules-based systems depend on constraining the language sufficiently that all possible language constructions and all possible meaning can be accommodated by the rules. As a result, the developer or team of developers that build the rules are in full control over the constrained language. Word embeddings mark a radical departure from this approach by using encodings that extract information from human language itself. In the cases of Word2Vec and GloVe this requires a large amount of text to produce representative models. For instance, the largest pre-trained Word2Vec model included in Gensim—an open source library for NLP tasks—was trained on a corpus built from Google News data containing about 100 billion words, and the various pre-trained GloVe models in the same library were trained on Wikipedia dumps—or copies of all of the content on English Wikipedia—containing about 6 billion tokens (Řehůřek & Sojka, 2010). From these datasets, both models attempt to represent the meaning of words in vector space. This means that they provide a way of encoding natural language into numerical values that computers can process, and since those values capture some semantic value, the computations are done on the meaning instead of just the characters.

One question that I will discuss at length in the fourth chapter that is of great interest in the public sector and academia is the question of intellectual property. In late 2023, *The New York Times* brought a case against Microsoft and OpenAI about the use of content scraped from news articles for training of LLMs (Grynbaum & Mac, 2023). Likewise, as I discussed in the previous chapter, plagiarism is a major concern for academics when considering the use of GenAI. These are interesting and important questions to ask, and I will address them, but what should become apparent in the discussion of word embeddings above is that the way that language models contain information from the data upon which they were trained is more complicated than copying from a source on the internet. Certainly, the word embeddings are derived from human language that was collected in the wild, but each word embedding is derived from numerous sources since the word may appear in numerous sources across the corpus. Likewise, it is likely impossible to reverse engineer how exactly the word embeddings were extracted because in the case of models that use neural networks, like Word2Vec and Transformer models, those calculations occur in hidden layers of the neural network and are blackboxed from the data scientists and researchers themselves. I will discuss ways we might think of plagiarism and intellectual property in the fourth chapter, but what is clear from word embeddings is that it is possible to move from a constrained rules-based approach to language computing and to encoded representations of language that have captured the meaning of language, as well. In the next section, we will arrive at Transformers, which are the models at the core of contemporary GenAI, and we will see that while they are considerably more sophisticated in how they encode language, they rely on similar strategies and are the next step from word embedding models like Word2Vec and GloVe.

3.5 Transformers

Word2Vec and GloVe both encode meaning of words based on their relationship with their immediate context or with the words that are statistically likely to appear with them. This was a major step in dealing with the language problem because it provided a way to encode meaning and not just characters as rules-based language algorithms might. However, while word embeddings do capture meaning, Word2Vec and GloVe word embeddings treat language more as a collection of words whereas humans encounter text as a series of relationships with many words together. In other words, while a word embedding might have 300 dimensions—meaning 300 features of the word have been captured in the vector—this still fails to capture the complexity of human language, and particularly human language beyond individual words. To accomplish the kind of encoding of language necessary for GenAI to generate texts that could pass for natural language, much more information needs to be represented in the model. The introduction of the attention mechanism changed this, and Transformers like GPT rely on this invention.

Attention mechanisms were first introduced in computer vision and quickly spread to other tasks that had previously been done using recurrent neural networks (RNNs) for improved performance and their ability to handle larger input windows (Brauwers & Frasincar, 2023). To understand Transformers, it is useful to begin with neural networks, and Andrew Ng (Machine Learning Simplified, 2017), who is perhaps one of the leading researchers and pedagogues in ML, provides an excellent overview in his course on Machine Learning. Neural networks are mathematical models built around "neurons," which are designed to simulate a neuron in the brain. Each neuron performs a series of mathematical operations on a set of inputs and a set of weights—or values to multiply these inputs by—and returns the output. These neurons are assembled in networks with multiple layers where the combined outputs of each layer are provided as inputs to the next layer of neurons. Earlier, I described a function that could be trained to model grades as dependent on hours of study, and for our purposes here it is sufficient to say that the particular mathematical operations that occur in each neuron are similar. We might consider each neuron to represent some feature of the data that is being input. When a neural network is trained, the weights contained in the neurons are automatically adjusted by feeding large amounts of data through the network and comparing the actual output to a target output until the distance of the output of all the values is as small as possible from the corresponding target output, in a similar way that we might adjust the coefficients in the study hours and grades function so the line graphed by the function is as close to all of the points in the dataset as possible.

There are a few different ways this training can happen, but RNNs require data to be fed through the system one element at a time. For each new element input into the system, the result of the previous sequence of calculation is also fed back into the system. So, if an RNN is being trained on language data, it must wait to process each word until the previous word has been processed. As a result, the encoding window cannot be particularly large as we can see with Word2Vec or GloVe, and more importantly the calculations cannot be done in parallel. Each word must be processed in sequence. Processing is limited by the processor cycles of the central processing unit (CPU) or graphical processing unit (GPU) in the computer. With an RNN, the limit, then, is the time it takes to run a processor cycle. Parallel processing would mean that the task of running calculations could be spread across a number of CPUs or GPUs (or more precisely cores within a processor). While having a number of processors running in parallel might be considerably more expensive per hour, it allows the training of a neural network to be done in less time. Attention mechanisms solve both of these problems by allowing each word to be processed independently of the preceding and following words.

Attention mechanisms simulate more closely how humans process language in that a larger span of language contributes to the interpretation of each individual words—perhaps the entire sentence or two sentences. The model produced by training with attention mechanisms comes much closer to representing language as relationships between words across sentences rather than just the words themselves. Since GenAI is made possible by LLMs that are built with the Transformer architecture, which, in turn, depend on attention mechanisms, it is worth unpacking these. The attention mechanism involves three components: the query word, keys, and values. The query word is the word that is being encoded. The keys and values represent all of the words in the sentence or several sentences. Keys represent the weight of each individual word, and the value represents the word itself. Each of these components is represented as a vector. The keys are weighted by assigning them attention scores based on their relationship to the query word. I will go into detail in this calculation below. There are numerous ways that feature vectors can be initialized for each word including the output of an RNN, like Word-2Vec embeddings, and there are a number of ways that the attention scores can be assigned (Brauwers & Frasincar, 2023).

Once the weights of the keys are calculated, they are used to calculate the weighted sum of all of the values, and the resulting vector is the encoding of the query word. While attention mechanisms still rely on ML techniques that automatically adjusting the weights until the output approaches the expected values, and while the values are still only mathematical representations of words, the attention mechanism comes closer to human processing of language because the information extracted from the relationships within language across larger spans of text—frequently a complete idea since the span will cover one or more sentences. Moreover, since each query word can be calculated independently without having to wait for the word before the query word to be calculated, the calculations can be done in parallel.

In addition to this advance in algorithms, hardware improvements have also made this training easier. The calculations involved in training models involve finding dot products and performing matrix multiplication, both of which require numerous individual calculations. When done on a CPU, this is a fairly slow process, but modern GPU's have allowed these calculations to be run much more rapidly and efficiently. GPUs happen to be designed for such calculations because they were designed to control the output of an array of pixels in a computer display simultaneously and in rapid succession. GPUs might have many processor cores and so they can conduct these calculations simultaneously. So not only do attention mechanisms allow for parallel processing but modern GPUs are designed particularly for these kinds of calculations. The major breakthrough with this new hardware and software appeared in 2017. Prior to that, attention mechanisms were often paired with RNNs, and so the limitation to parallel processing remained. In 2017, researchers at Google and the University of Toronto introduced the Transformer architecture, which abandoned the RNN completely and used only attention mechanism yielding a model that could do machine translation of natural language that could perform as well or better than other models and that could be trained in a fraction of the time (Vaswani et al., 2017). Authors Vaswani et al. suggest that this approach can "reach a new state of the art in translation quality after being trained for as little as twelve hours on eight P100 GPUs" (Vaswani et al., 2017, p. 2). The training cost for training models to do English to French translations using their Transformer was one to two orders of magnitude less than training previous systems, as well (Vaswani et al., 2017, p. 8). This increased performance and lower cost stems from removing the RNN and the resulting increased parallelization possible with the Transformer architecture. The development of Transformers and parallelized processing corresponds to breakthroughs in increased accessibility to GPU instances in cloud computing providers.

Transformers are built around attention mechanism and eliminate the need for an RNN, and in reality, they are trained through a series of steps. First, the training data is preprocessed and encoded into word embeddings. These might be those generated by Word2Vec or GloVe models, or they might be encoded with vectors containing random values. After the input data is encoded, the language is processed through a number of layers of multi-headed attention mechanisms. Attention refers to the ability for the calculation to accommodate not only the query word, but also the keys and values for all of the words in the same sentence as or several sentences surrounding the query word. Multi-headed means that query words, keys, and values are "project[ed] … *h* times with different linear projections" (Vaswani et al., 2017, p. 4). For each head, the values of the vectors are modified by processing them through a different linear function. The multi-headed architecture was developed because researchers found that it yielded better results. As query word, keys, and values move through the system, the attention function adjusts the values of the keys (or the weights), performs additional calculations, and returns an output vector that represents the word in the context of the sentence. Like other language modeling approaches that rely on ML, the Transformer model is "trained," meaning the various weights are adjusted based on patterns that appear in the training dataset. Since each head uses a different linear transformation, different patterns in the dataset are applied to the weights as the system runs, and each of these patterns that is captured in the value of the weights represents some characteristic or feature of the language. For instance, one transformation function might represent the positional relationship between words, and another might represent the dictionary definition of the word and so on. However, since the training is unsupervised, and the dataset is not labeled, the Transformers extract patterns that appear naturally in the mathematical representation of the data. As a result, the training dataset must be quite large for interesting patterns to emerge. However, if one were to inspect the transformation occurring in any of the attention heads, it would not be possible to identify what characteristic of language the head was representing only that it was representing some feature that does exist in the dataset. This will be interesting when we discuss the generic output of inference using Transformer models in the next chapter because while we may not know exactly what features is being modeled, that feature is reproduced in the output of the GenAI system. After each head has processed the query word, keys, and values, the resulting vectors are combined yielding a final word embedding that provides a multifaceted representation of the various relationships between words, the sentences they appear in, and the other features of language that have been extracted. Again, since GPT-3.0, these models have been able to extract a sufficient number of characteristics of language that they can be used effectively to perform realistic NLG that we see in GenAI systems today.

Since Transformers must be trained on staggeringly large datasets, they still remain expensive and time consuming to train—even if they are less expensive and faster than RNNs—and as a result most GenAI systems use pre-trained Transformers. There are a number of different types of Transformers that can be used in GenAI. ChatGPT uses the GPT model, which is a general LLM trained on publicly available language. OpenAI offers several other models as well, and each of these models have strengths and weaknesses. One popular model is the Bidirectional Encoder Representations from Transformers (BERT) model, which uses the Transformer architecture and is first pre-trained on two tasks: predicting masked tokens within input sentences and predicting the next sentence (Devlin et al., 2019). Based on the correctness of each prediction it adjusts the weights of each vector automatically. The model can then be fine-tuned according to the task for which the resulting model will be used. As is the case with most pre-trained models, the fine-tuning step makes a significant difference but is also relatively inexpensive because a small amount of data and training time is needed. Experiments find that BERT models perform considerably better than traditional NLP approaches to tasks like text classification, sentiment analysis, and others (Devlin et al., 2019; González-Carvajal & Garrido-Merchán, 2023; Tenney et al., 2019). In addition to Masked Language Models like BERT, there are other Transformer models including those that are primarily encoders, primarily decoders, or contain both. The encoder converts input data into a contextual representation, which is particularly useful for generating an "understanding" of the language. The decoder generates output data using the attention mechanisms, which is particularly useful for creative tasks like poetry, story-telling, and visual art. Finally, encoder-decoder transformers use the encoder component to convert input data into contextual representation and the decoder data to generate outputs based on that contextual representation. ChatGPT uses a primarily decoder architecture is quite effective at generating output, but requires extremely specific prompts to do so (Kocoń et al., 2023). The original Transformer model presented in "Attention is All You Need" is an encoder-decoder model that uses attention mechanisms for both encoding of input data and decoding of output (Vaswani et al., 2017).

So far, I have not extensively discussed the nature of the massive datasets on which transformer models have been trained, and there are serious ethical concerns regarding these data and their collection. Much of the current literature and nearly all of the current position statements on GenAI emphasize the ethical concerns; however, I believe it is critical to understand exactly what these systems are and how they've used the data before we begin to discuss the ethical concerns, and so I will turn to those in the next chapter. At this point, LLMs are built on Transformer architecture. They are trained using attention mechanisms, meaning the vectors that represent language in these models are iteratively modified by automatically adjusting the weights or numerical values associated with various parameters associated with the larger context—usually a sentence or two—in which each word exists in the dataset. I earlier discussed modeling of grades versus hour studied. That model contains a single parameter, hours studied, and can be represented in two-dimensional space. These large language models might contain many billions of parameters and therefore represent many dimensions of language. As the models are trained, the parameters are adjusted until final weights are established, and the model can then be used either for further fine tuning with additional context specific data or for inference which involves performing some NLP or NLG task like sentiment analysis, topic modeling, or even generating human-like natural language.

Once these models have been trained, they can be used for NLP and NLG tasks. This process is called inference, and in inference input data is fed into an inference pipeline where it is first preprocessed, input into a transformer architecture, once the input is processed, output is predicted one token at a time. So when I prompt ChatGPT or Gemini to provide a summary of Rhetorical Genre Theory, it takes my prompt, converts it into vectors which then match with the vectors in the LLM that represent the data from the training set on Rhetorical Genre Theory, which likely included blog posts, publicly available articles, and other online texts that summarize theory, all of which were certainly scraped and included in the dataset on which the LLM was trained. Once this region of the LLM has been identified, the inference pipeline predicts token-by-token—or semantic unit by semantic unit—the most probable token matching the output of the prompt. The output is likely post-processed for grammatical correctness and often accuracy (Google Gemini allows users to see other drafts, which means that the system generates multiple outputs and then re-processes those outputs again to find the one that matches best). The significant limitation to LLMs is that while they can produce language that seems very much like natural language, they cannot generate something that is not part of the training set. I will discuss this further in the next chapter as I analyze the responses to a number of prompts to position GenAI in our larger understanding of language, composition, and rhetoric.

3.6 Generative AI

Throughout this chapter, I have provided a survey of how Computer Scientists, Computational Linguists, and Data Scientists have gone about solving the language problem. It would be worthwhile for scholars in Writing Studies and Education to further explore this field. However, for now, I suggest that if we are to engage with GenAI, and especially, if we are to make effective recommendations on its use, it is vital that we understand how researchers have discovered ways to solve the language problem and model language in ways that allow computers to process it. This begins with understanding how computational systems were designed and work. Computer Science uses *computational thinking*, a disposition containing a number of heuristics that break down human problems into smaller ones that can be solved by technology. For the problem of language, rules-based systems solve this by constraining language to the point that it can be contained within a set of prescribed instructions for the computer. Such systems are not really working with "natural language," but they do allow humans to interact with them through language that is more akin to natural language than a programming language might be. As hardware and ML techniques advanced through a considerable amount of trial and error and experimentation, NLP and NLG become more feasible. The first systems relied on human labeling and hybrid approaches where natural language could be provided as an input and preprogrammed responses would be returned. Additionally, word embeddings made it possible to perform tasks like sentiment analysis, topic modeling, and other tasks within Computational Linguistics. However, many of these systems depend on labeling and so their ability to process language depends on the constraints of the labeling process, which involves hours of human labor applying various labels to data that can be fed into supervised ML systems.

Modern NLP and NLG systems have made using unsupervised learning more feasible. These systems extract patterns from massive corpora of data using ML techniques. Word embedding systems extract and encode features based on individual words and their local and global contexts, and they treat language as a collection of words. The contemporary breakthroughs that allow GenAI to be possible, and as effective as they are, depend on models that encode the relationships between words in sentences. Language is treated as a collection of relationships that traverse single or multiple sentences. These relationships are extracted from performing numerous analyzes of truly enormous datasets, yielding models that, while they still treat language as mathematical operations, simulate very closely natural human language. The fact that LLMs work strongly reinforce structuralist views of language (Gastaldi, 2021). LLMs are trained by allowing transformer systems to identify naturally occurring patterns across massive datasets. This suggests that there are naturally occurring patterns in language, suggesting that language itself contains some reliable structure that gives meaning to words. Since this project is not focusing on a philosophy of language, but rather is attempting to theorize how we might understand GenAI in education, composition, and rhetoric, I will not be pursuing this topic, but it would be a highly fruitful avenue for future research. At this point, hopefully, we have developed a greater understanding of how LLMs model and generate text.

Note

1 There are NLP processes that focus on phonetics and phonology (Boersma et al., 2020).

4 Writing Theory for Generative AI

4.1 Writing Studies and GenAI

In the previous chapter, I provided a survey of the approaches Computer Scientists, Computational Linguists, and Data Scientists have taken to address the challenge of processing human language with computers. Over time, they have built computational models that have encoded natural language into numerical values in various ways that allow a computer can interact with it. The current breakthroughs in GenAI depend largely on a convergence of the Transformer architecture and advances in GPUs. One of the major challenges that helps us position GenAI within our understanding of rhetoric is that it is quite difficult to capture enough information about human language to represent it in a computational model. Earlier neural networks that could do unsupervised learning and extract information from the language automatically required considerably too long to train, and supervised training required a great deal of human time and was limited to how researchers labeled data. The advances made possible by Transformer architecture means that LLMs can contain sufficient amounts of information that they can begin to represent natural language effectively. Information is captured in the weights or the parameters of the model, and generally more parameters, the more information can be modeled.

There has been a progression in how much information could be extracted from language from grammar to semantics to rhetoric. Early transformer models like GPT-1 had fewer parameters, and therefore the main feature of language it captured was grammar. As a result, NLG systems using this model could produce grammatically correct text, but frequently it did not make much sense—or it was not particularly meaningful. Around GPT-3 and especially GPT-3.5 the number of parameters and the architecture of the GenAI system became sufficiently large that these models seemed to be modeling the natural language fairly well, and they seem to have encoded rhetorical knowledge (Cress & Kimmerle, 2023, p. 611). As LLMs advance, it is likely that we will find that the ability of GenAI systems will continue to improve as they are able to more effectively encode language features and retrieve them in a way

DOI: 10.4324/9781003493563-4

that generates increasingly more appropriate text. Now that we have a better sense of how that information was encoded and how text is generated, we can begin to discuss rhetorical theory and the way that scholarship has positioned GenAI within the context of writing and other rhetorical activities. In this section, I will discuss existing theory and spend some time pushing back against the metaphor of "collaboration." In the next two sections, I will offer some ways we might think of GenAI as it relates to Rhetorical Genre Studies (RGS) and Post-human/network theories of rhetoric. I select these theoretical frameworks because RGS provides a useful way to think about the conventions of writing to which GenAI seems to effectively conform, and network theories of rhetoric are particularly useful in revealing the interactions that occur across complex systems of which the GenAI pipeline is a part.

The majority of the writing on GenAI and LLMs in the field of Writing Studies has appeared since the release of ChatGPT-3.5, and most of the scholarship focuses on the practical applications of GenAI systems like ChatGPT and pedagogical implications in our classrooms. Much of this scholarship has offered excellent recommendations, particularly when these recommendations focus on guiding students to use GenAI within their writing process (Byrd & Flores, 2023; Cress & Kimmerle, 2023; Dobrin, 2023). However, there is considerable room to further theorize LLMs and GenAI in the context of Rhetorical and Writing Studies theory. In her review of the literature, Selena Sampson Anderson (2023) suggests that GenAI is usually described using one of two metaphors: tool or collaborator. As a tool, GenAI is something that writers might engage with and use as part of their writing process. This metaphor provides little agency to the system, which is appropriate. However, if we see GenAI as a tool, it is important to understand how it works, which was my aim in the previous chapter, and how it fits into our larger understandings of rhetoric. The metaphor of collaboration is less appropriate because it anthropomorphizes GenAI systems in inaccurate ways and more importantly it presents a risk to student agency. In spite of viable alternatives like the compelling metaphor of "co-constructing" (Cress & Kimmerle, 2023), the "collaboration" metaphor is becoming commonplace (Anderson, 2023; Cummings et al., 2024; Daniel et al., 2023; Fui-Hoon Nah et al., 2023; Kotzeva & Anders, 2023). This suggests that we are thinking of GenAI as something like human even if we simultaneously acknowledge that the systems are generating text through statistics and probability. It is important that we remember that GenAI is a system that includes a number of algorithms and models collected together in a pipeline of interrelated processes, and it is not sentient, thinking, or even aware of the meaning of the prompt in any sort of human way. GenAI is an impressive technology, but it is intelligent only in the sense that intelligent algorithms can make real-time decisions based on incomplete data (Wang, 2019). Perhaps in the future—and even the near future—we will have "thinking" and "understanding" AI, but for now, this remains in the domain of science fiction.

It is no surprise that the collaboration metaphor would be so compelling. Without looking into the complex, multilayered system that behind it, most users experience GenAI through what they can see in the UI. Numerous GenAI systems like ChatGPT or Gemini or Claude give the user a chat-like UI and a user experience that is uncannily similar to chatting with a human. It makes sense that the companies would provide this sort of interface because it certainly appeals to the public imagination, and more importantly, it gives users a very familiar context of genres.

This notion of the appeal by a familiar context and set of rhetorical actions can be explained effectively by RGS. Research in this area has revealed that our interpretation of rhetorical situation and rhetorical action is strongly influenced by prior genre knowledge (Bawarshi, 2003; Bazerman, 2003; Reiff & Bawarshi, 2011). In other words, when we encounter a rhetorical situation, we draw on our prior genre knowledge to understand the situation and frame an appropriate response. This process is often referred to by knowledge transfer researchers as "uptake" (Dryer, 2016; Freadman, 2002; Medina, 2018; Rounsaville, 2012, 2014). In the context of education, students read writing prompts created by teachers, and as they do so, their past experience with writing prompts influences how they understand the present prompt. Since RGS borrows the epistemological and ontological perspectives that are present in Foucault's theory of discourse, it is fair to say that the structure of genre itself contributes to the individual's understanding of the world around them—particularly the rhetorical situation—and also their identity position therein (Bawarshi, 2000; Foucault, 2002). Stated more simply, our past communication experiences with genres define how we understand and participate in present and future ones.

This perspective from RGS provides us with significant insight into our uptake of GenAI. It is safe to assume that most users of GenAI have experienced chatting with another human via a text-based chat application. They probably have spent enough time using this kind of application that performing the associated genres have become naturalized, and therefore invisible. RGS would suggest that most users of GenAI, before they even know what this technology is, will already know how to interpret the way to interact with the UI of an instant-messenger like application and will already have experience *being* in that space because they have already internalized the genre. In other words, they know how to do text-based chat and have the intuitive sense that the voice on the other side is that of another person. The companies have designed the UI of their GenAI chatbots to fit these genre expectations quite well. At the time of writing, the main pane of ChatGPT's user interface is a text area with convenient suggestions for prompts including "Message to comfort a friend," "Plan a relaxing day," "Activities to make friends in new city," and "Overcome procrastination" (*ChatGPT*, 2024). At the bottom of the pane is a text entry field with the placeholder text "Message ChatGPT." This placeholder text suggests that ChatGPT is something that can be a recipient of messages, which in our experience would be a human entity. Gemini provides a similar user interface. In the main pane of the interface, the background says "Hello. / How can I help you today?" (*Gemini—Chat to Supercharge Your Ideas,* 2024). This interface also offers prompt suggestions that can be clicked, and a text entry bar with the placeholder text "Enter a prompt here." Once again, the user interface using the first-person singular pronoun positions the system as a human or human-like entity. Claude goes even further introducing itself when a user signs up for their first chat. It says, "Hello, I'm Claude. I'm a next generation AI assistant built for work and trained to be safe, accurate, and secure. I'd love for us to get to know each other a bit better," and it prompts the users to reply: "Nice to meet you, I'm…" followed by the placeholder text "Enter your full name" (*Claude*, 2024). The user interface of all of these GenAI systems try to appear to be like a chat partner. On the one hand, each of these systems does clearly identify that they are GeneAI. However, the interpretive, constituting lens of genre is already at work when a user sits down to "chat." In other words, genre begins situating us in rhetorical space prior to our conscious awareness of that space. This means that even though the user is consciously aware of the artificial nature of these chats, their experience—or uptake—of "chatting" with GenAI will be always-already informed by their prior experience of the genre of text-based chats with other humans. Certainly, there are sufficient opportunities for the user to recognize that they are not talking to other intelligent being, but it definitely benefits the companies offering GenAI services for people to anthropomorphize their systems. Further research into the effects of this anthropomorphizing on the users would be particularly useful.

I suggest that collaboration is such a problematic metaphor because rhetorical action is a balance between constraints of what is recognizable and choice, and while that "choice" can be made by the algorithms within the GenAI system and LLMs, the agency that an individual might have is in that choice, and one of the main reasons we would study and teach writing or rhetoric is to improve students ability to affect the world around them through their rhetorical choices (Devitt, 2004; Nowacek, 2011). If we consider AI a collaborator, we risk inappropriately relinquishing some of that choice and agency to that system. If we teach writers to think of GenAI as a collaborator, we encourage them to turn over some of their own agency in the writing situation to the system.

The notion of writerly agency is important because it emphasizes the verb-ness of the word writing and collaboration fits into this writing-as-action perspective. Writing is an action that is done to produce a material response to a rhetorical situation. Scholars in RGS, since its inception, have argued that genres are actions that we perform in response to rhetorical situations to the extent that it has become a feature of genre research to begin by citing Carolyn Miller's article that claims this (Freadman, 2012; Miller, 1984). A more dynamic view of genre suggests that they should be "[understood] as both constraint and choice, both regularity and chaos, both inhibiting and enabling" and when writers understand genres this way they can "use the power of genres critically and effectively. In such power is individual freedom" (Devitt, 2004, p. 156). Devitt's emphasis is on the pedagogical potential of genre, but there are larger implications of this as a way to understand how people perform writing. Consider a person going on the job market, and they find a job posting for company A. That person will read and interpret the posting through the lens of their prior experience with job postings. In doing so, they find themselves in a rhetorical situation that demands that they put together an application.

The nature of rhetorical situations has been debated by numerous scholars, but the consensus is that we use rhetoric in response to situations that compel us to do so (Biesecker, 1989; Bitzer, 1968; Consigny, 1974; Miller, 2024; Vatz, 1973). The person looking at the job posting feels an exigency that compels them to apply because they need a job, and they identify this particular job as one for which they are qualified. Within the rhetorical situation, the applicant will recall the genre knowledge they have internalized and likely identify that a cover letter and a resume will be the most appropriate way to respond. While these potential genres that the applicant has identified may seem stable and conventional, there is considerable play that these genres allow. Ideally, the applicant has done research on company A, discovered and perhaps met people who have worked there, and maybe even done an informational interview with the company. Since job postings tend to be quite generic, the applicant has probably gone beyond the posting and done additional research to identify what are the particulars of the job itself. Based on all of this knowledge, insight and awareness about the specific rhetorical situation, the person will make choices about how to put their resume and cover letter together. Some of these choices are constrained by the genre conventions, but some of these choices will be based on the feeling and intuition of the individual, and they might even choose to intentionally violate some of the conventions of the genre. The point being this feeling and intuition is a crucial component of the writing process and allows for the individual's response to be unique and novel. It is precisely this interplay between regularity and chaos that makes human language difficult for computers and precisely where GenAI systems fail, even if the language models have encoded some genre knowledge.

RGS suggests that writing involves a delicate negotiation that requires not only knowledge of conventions and experience with prior genres, but also and as important, it requires awareness of the rhetorical situation and potential room for improvisation that may exist in *this* particular rhetorical situation. When collaboration is introduced, two or more people work together to respond to the rhetorical situation. When I collaborate with another writer, the process is a negotiation in which we together assess the rhetorical situation, ascertain the collective rhetorical resources we have at our disposal, and negotiate the most appropriate response based on each of our conscious and embodied perspectives. Each member of the collaboration has both conventional genre knowledge, but more importantly intuitive awareness of the rhetorical situation. The virtue of collaboration, in my mind, is that if the members collaborate effectively, they can mutually gain a perspective of the rhetorical situation that contains both multiple sets of genre knowledge, but more importantly, multiple intuitive and subjective points of view. While contemporary GenAI systems might have encoded considerable rhetorical and genre knowledge about past human language, they do not have awareness of the present situation because they are not aware in any human way, and they particularly do not have any capacity to identify potentials for improvisation that are so crucial to rhetorical actions. This is not to say that GenAI is not something that can help writers better understand and respond to rhetorical situations, but it is vital that we do not overstate the power of these systems to do language. At this point, they are not doing language, but rather giving us access to patterns that exist in language via a human-like chat interface. I suggest that while they can give us insight into language, we perhaps should not see them as collaborators in a negotiation.

While GenAI cannot really participate in responding to a rhetorical situation in any human sense, using the metaphor of collaboration is particularly problematic because of how it positions the agency of the writer, especially student writers who are frequently always already positioned as recipients of knowledge and instruction rather than producers of useful work (Freire, 1968). Various movements within Writing Studies have expressed concern from writerly agency from the Expressivist Process movement's emphasis "voice" to contemporary movements focusing on Translingualism and antiracist pedagogy. Implied in the notion of "voice" is the goal of promoting writerly agency, and the metaphor of collaboration presents a threat to this. For instance, Fiona Fui-Hoon et al. argue for a "human-centered AI" (HCAI) or a "human-in-the-loop" approach to GenAI (2023, pp. 292–293). This approach is explained as "[h]umans and generative AI [collaborating] harmoniously, with humans at the center and AI functioning as assistants" (Fui-Hoon Nah et al., 2023, p. 293). The emphasis on centering the human participation in the writing is a fair argument, but even as assistants, the collaboration model gives some of the decision-making over to the GenAI system. Furthermore, the notion of centering humans in interactions implies that humans could be out of the loop, so to speak. GenAI does pose the threat of putting the writer out of the loop, particularly when it can effectively produce texts with considerably less work. This is precisely why it is vital that we take great care when employing metaphors like "collaboration."

Not only is the notion of collaboration problematic because it takes agency away from the human writer, but also it is problematic because it gives agency to the GenAI system that the system does not and cannot presently be an agent in any human sense of the word. The notion of agency has been a feature of research on ethics of bot rhetoric, but much of this work focused on earlier rule-based or intent-based bots that required considerable control by the developer (Coleman, 2018; Martin, 2019). Human agency in writing is in the ability to negotiate between the conventions and chaos of genres within a rhetorical situation (Devitt, 2004). GenAI merely generates output text that corresponds to the values of the input text. In other words, it purely follows the conventions of writing in the training data that correspond to the encoded input, and it lacks awareness of the situation to be able to make *these kinds* of rhetorical choices. Hallucinations are a good example of the limitation in considering GenAI a collaborator or writing agent. First, the term hallucination is perhaps problematic.[1] GenAI hallucination occurs when the system generates content that *is* there but is not correct to what the human user wanted. According to Jodie Burchell (2024), there are two types of hallucinations that are produced by inference pipelines using LLMs: hallucinations caused by faithfulness and hallucinations caused by factualness.

Hallucinations caused by faithfulness are those that are caused when the system generates text that is unrelated or inaccurate in relationship to the question. For instance, if I asked a GenAI system to name the largest mountain in North America and it named rivers, that would be a faithfulness hallucination. Something went wrong in the pipeline and either my prompt was bad, there was a problem converting my prompt, or there is a bug somewhere in the system. Despite the broken state, GenAI still generated output when it perhaps should not have. Anecdotally, this seems to be occurring less frequently, probably because this is a known issue that engineers have been attempting to solve.

Factual hallucinations occur when the LLM has been trained on incomplete or inaccurate data. When I ask ChatGPT, for instance, to write a bio about myself it writes:

> Dylan London Medina is a dynamic professional excelling in technology, creative arts, and social entrepreneurship. With a degree in Computer Science, he swiftly advanced as a software engineer, renowned for blending technical prowess with innovative solutions. Dylan is also a talented musician and visual artist, with his work showcased in various galleries and digital platforms. As a social entrepreneur, he founded a non-profit to provide underprivileged youth with access to technology and creative education. Dylan's diverse talents and compassionate leadership continue to inspire and empower others, making him a true modern Renaissance individual.
>
> (*ChatGPT*, 2024)

When it comes to the particular details, it is fiction, and ChatGPT says as much when I ask it, "Is this person a real person?" ChatGPT replies: "No, Dylan London Medina is a fictional character created for the purpose of the bio. If there happens to be a real person with that name and similar accomplishments, it would be purely coincidental" (*ChatGPT*, 2024). The LLM does not contain any information about me. However, since there was no information about me in the training data that could generate a factual response, the LLM positioned our discourse in the domain of fiction rather than non-fiction, and generated a conventional fictional bio—flattering, but alas, a work of fiction. In fact, calling it a hallucination is perhaps not correct. The GenAI system did not "invent" or "make up" something that it should have known. Instead, it identified my request as fiction because according to GPT-4o, Dylan London Medina is a fictional character. It then wrote a generic fictional bio that could fit nearly anyone.

However, when I ask it to write a 100-word biography for Joe Biden, ChatGPT writes:

> Joe Biden, the 46th President of the United States, is a seasoned statesman with decades of public service. Born in Scranton, Pennsylvania, and raised in Delaware, Biden's political journey began as a young senator, where he earned a reputation for bipartisanship and dedication to his constituents. Serving as Vice President under Barack Obama, Biden played a pivotal role in shaping domestic and foreign policy, notably leading efforts on healthcare reform and diplomacy. A unifying figure known for his empathy and resilience, President Biden continues to prioritize issues such as climate change, healthcare, and economic recovery, striving to build a more inclusive and prosperous future for all Americans.
>
> (*ChatGPT*, 2024)

This information is true and is not a hallucination. ChatGPT responds to the question about the reality of this bio saying, "Yes, Joe Biden is a real person. He served as the 46th President of the United States, having been inaugurated on January 20, 2021" (*ChatGPT*, 2024). "Served" is perhaps not appropriate since at time of writing Joe Biden is still serving, but ChatGPT's GPT-4o model was clearly trained on data that contains biographies of Joe Biden, and so it was able to generate a factual, generic bio about Biden.

These hallucinations, often simply addressed as a limitation of LLMs in the literature, provide useful insight into the GenAI system. Faithfulness hallucinations reveal the limited ability of GenAI to "understand" our prompts. If the input is not good, the output will be wrong. Factual hallucinations reveal information about the training data. A serious concern with factual hallucinations that Burchell (2024) discussed at her talk at PyConUS 2024 is when the training data contains misinformation or intentionally false information like conspiracy theories and propaganda. While most of the dominant GenAI systems were trained on datasets that were cleaned for content that could promote dangerous factual hallucinations stemming from misinformation, it is important to note that if the training data contained some idea, the GenAI system does not have any way to determine how "true" true the data is unless it exists in the context of untruth. One certainly could train an LLM on data and create a GenAI system that would only produce conspiracy theory nonsense. The point of all of this is that factual hallucinations are not actually representing something that is not in the model. Quite the contrary, factual hallucinations occur when what is in the model does not match what we expect. GenAI systems generate texts following the patterns of the language that they were trained upon. This derivation might have some degree of variability—a variable called "temperature" allows developers to control how much error can exist in the calculations to generate the output text—but it is still a derivative of language that was already written. It is for this reason that even considering GenAI an assistant or a co-constructor of knowledge could be problematic (Cress & Kimmerle, 2023). If we believe that GenAI has agency and can participate in the construction of knowledge, we are misunderstanding how the system works. However, once we set aside notions of collaboration with GenAI and resist the urge to offload rhetorical agency onto the system, there is significant potential for using this technology. Since the LLMs have so effectively encoded features of natural language, we can gain substantial insights into that language by using GenAI.

4.2 Encoded Genre Knowledge

Since GenAI is so capable at generating highly conventional texts thanks to the fact that the LLMs that they incorporate were trained on data from the conventional, dominant language, they fit well into the framework that RGS provides for thinking about the conventions modeled in LLMs. As I discussed above, RGS scholars have discussed the balance between conventionality and novelty of rhetoric, and I suggest we might see genre as patterns of rhetorical action that exists within discourse. Above, I emphasized the action aspect of genre and the opportunity for creativity that they provide, but other scholars have emphasize the stability of genres and its ability to allow us to understand and interpret texts, and identify and respond to rhetorical situations (Bazerman, 2003; Miller, 1984; Swales, 1990). In this case, the important aspect of genre is as a sort of heuristic or mental model that helps its users interpret a rhetorical action or situation. Bazerman calls genre a "psycho-social recognition phenomena" (Bazerman, 2003, p. 317). By this, he means that genres exist in the intersection between the social or discursive and the psyche of the individual, and they are used for recognition. In other words, genres exist in a discourse, and they are internalized by the individual member of that discourse. Once internalized, the individual can use genres to identify, interpret and respond to rhetoric and rhetorical situations.

Consider the writing prompt that teachers produce for their students. Teachers find themselves in situations in front of students where they have to assign work, and since their process of becoming a teacher has taught them how to do this, they have internalized a set of appropriate genres to guide how they do this. One such genre is the writing prompt, which needs to first tell students what subject they are to write on either by theme, theoretical lens, or perhaps with a particular question to be answered. Second, it needs to give students the formal constraints of the writing task. Third, it probably should give students some insight into why they are being asked to do this task and how the task will be assessed. This genre knowledge serves as a sort of heuristic that guides the teacher as they begin to construct their writing prompt which they will send to students. If the prompt is successful, the students will, for the most part, be persuaded to do the particular type of work the teacher asks. To assess the genre knowledge that may be encoded into the LLM, I have prompted ChatGPT and Gemini to help me respond to such a rhetorical situation. Since each of these interactions what quite lengthy, I'll be citing sections from them, and including QR codes which will take you to the conversation. The URL links can also be found in the references section.

Beginning with ChatGPT using the GPT-4o model, I asked, "I want my first-year composition students to write a genre analysis of protest signs from some social movement of their choice. How should I do this?" My question seeks to see what sort of explicit genre knowledge is encoded in ChatGPT. It responds quite well, providing "a step-by-step plan to guide your students in writing a genre analysis of protests signs from a social movement" (Open AI, 2024). The five steps include "Introduction to Genre Analysis" in which I begin with a "Lecture/Discussion" and "Provide examples of genre analyses for different types of texts" with the goal to "Ensure students understand what genre analysis is and its purpose" (Open AI, 2024). The second step is "Selecting a Social Movement" in which the teacher "Help[s] students choose a social movement" (Open AI, 2024). ChatGPT suggests that this is done with a "Brainstorming Session," which is a discussion of social movements, and with "Preliminary Research" (Open AI, 2024). After this, students need to do the step called "Collecting Data" in which students are directed to "online archives, social media platforms, news websites, and other sources." An interesting piece in this step is that ChatGPT suggests that students are explicitly instructed on how to select sources: "Encourage students to select a diverse range of signs in terms of message, design, and context." This is an interesting point because one of the challenges of moving students into more complex writing is to encourage them to look at a broader variety of data. As they account for this array of data, their writing will need to become more complex to accommodate the internal complexities of the data. Once students have collected data, they should do "Analyzing the Signs" focusing on "Visual and Textual Elements," "Rhetorical Strategies," and "Context." These are the important components of a very generic genre analysis, and I would argue that there are more interesting things that can be done with this analytical lens, but ChatGPT is not wrong. The final two steps involve "Writing the Genre Analysis" and an optional "Presentation." In the writing section, ChatGPT provides a fairly conventional "Process" approach to writing moving from developing a thesis, to writing an outline, to drafting, to peer review, and finally revisions (Open AI, 2024).

https://chatgpt.com/share/22a4faac-6fac-451b-bfe5-a8c2ef237e70

Figure 4.1 Scan to access ChatGPT Conversation: "Genre Analysis of Protest Signs".

I followed up ChatGPT's response with the prompt "Can you draft the thing I'll give to students when I assign this task?" I intentionally used fairly vague language and since these chat interfaces use a parameter in the function that allows the conversation history to be input back into the NLG pipeline so that that data is included in the calculation of future turns, "this task" is unambiguous to the system. ChatGPT responds with what I would argue is a very conventional writing prompt. The prompt begins with an "Overview" section that in three sentences describes what the students are asked to do. This is followed by a list of four "Objectives," which include "Understand and apply genre analysis techniques," "Explore the rhetorical strategies used in protest signs," "Analyze the interplay of visual and textual elements," and "Contextualize protest signs within their historical and cultural framework." After providing a brief overview and objectives, which are key features of the writing prompt genre, ChatGPT's writing prompt gives students five steps that correspond with the five steps it recommended in the previous response. It concludes with formal details ("5-7 pages…Typed, double-spaced, 12-point Times New Roman font, 1-inch margins") and a breakdown of how the grade will be calculated. The last section is a short "Tips for Success" list that are the sorts of hints a teacher might include at the end of a writing prompt. I also followed up this response by asking ChatGPT to "create the grading rubric where each criterion has 4 levels from needs improvement to exceeds expectations." Again, ChatGPT was able to generate a conventional grading rubric. I have included one of the sections below for "Understanding Gerne Analysis" which is worth 20% of the grade:

Considering how little information I provided ChatGPT, it was able to effectively teach me how to teach my students genre analysis, and if I had little experience with RGS and teaching students this skill, it would give me a reasonably good place to start. However, the lesson plans, writing prompt, and

Table 4.1 Rubric Generated by ChatGPT

Level	*Description*
Exceeds expectations (18–20)	Demonstrates a thorough and nuanced understanding of genre analysis. Accurately identifies and explains the genre's features, conventions, and rhetorical strategies.
Meets expectations (14–17)	Shows a clear understanding of genre analysis. Identifies and explains most features, conventions, and rhetorical strategies.
Approaching expectations (10–13)	Demonstrates a basic understanding of genre analysis. Identifies some features, conventions, and rhetorical strategies but lacks depth.
Needs improvement (0–9)	Shows limited understanding of genre analysis. Fails to accurately identify and explain key features, conventions, and rhetorical strategies.

https://g.co/gemini/share/5b2f1eed1a5f

Figure 4.2 Scan to access Gemini Conversation: "Analyzing Protest Signs".

grading rubric are conventional, and there are certainly other ways we could teach genre analysis that are more interesting. This is not a surprise as GPT-4o has been trained on data that certainly includes both instructions on how to teach genre analysis and writing prompts and rubrics for assigning student the task of completing a genre analysis. Clearly some genre sort of genre knowledge and genre awareness has been encoded into the model.

I prompted Gemini with the same prompts as I did ChatGPT, and Gemini's responses are surprisingly quite different from ChatGPT. First, the response from Gemini is considerably shorter. It focuses on the immediate activity of genre analysis rather than the sequence as ChatGPT did. Gemini's response breaks the genre analysis process into three sections: "Preparation," "Analysis Steps," and "Writing the Analysis" (Google, 2024b). "Preparation" involves selecting a movement and protest signs to analyze, and in the examples of social movements it does not include "Women's Rights" or "Black Lives Matter," both of which were examples in ChatGPT. It would be a worthwhile analysis to compare different platforms and different LLMs to investigate how they have encoded history of social movements. Another major difference is the steps of analysis that Gemini recommends. In the "Analysis Steps" section there is no explicit mention of "rhetoric" as there was with ChatGPT's response. However, Gemini's response does guide teachers to have students do rhetorical analysis even if it does not call it that by asking students to analyze the "Purpose and Audience" and "Effectiveness." The "Purpose and Audience" section suggests that the teacher "Ask students to consider the goals of the signs. What do they aim to achieve? Who is the target audience (authorities, bystanders, media)? How do the signs tailor their message to this audience?" In the "Effectiveness" section, the response says, "Have students evaluate the effectiveness of signs based on identified conventions." Gemini also provides a writing prompt, which begins with an overview explaining in two sentences what students will do. Then it provides details about the task that follow the pattern of the previous response. Instead of a rubric, the prompt includes a list of features for "Writing Analysis" or how the teacher will analyze the students' writing. This prompt also concludes with tips and the last paragraph explains why students are doing this assignment: "This assignment will help you develop critical thinking skills, hone your visual literacy, and gain a deeper appreciation for the power of communication in social movements." While Gemini did not provide an explicit rubric for grading, I did prompt it to create one, which it did:

The rubric is very much like the numerous rubrics I have encountered in my experience in the field, which is no surprise because such assignments and rubrics likely appeared in the training data used to create Gemini's LLM (Google, 2024b).

ChatGPT and Gemini can both generate instructions to teachers on how to teach genre analysis and generate writing prompts and rubrics that, with only a little revision, could be given to students. The content that these GenAI systems created is quite conventional, which makes sense. The question we might ask is does GenAI contain Genre Knowledge? While a human teacher with some understanding of RGS could definitely produce more interesting and more effective assignments, and a human teacher could design their assignments based on their knowledge of the particular class of students—something to which the GenAI system does not have access, both ChatGPT and Gemini are able to produce content that conforms to the genre of genre analysis writing prompt. This means that enough examples of genre analysis writing prompts existed in the training data that the features of this particular genre were encoded into the LLM to a sufficient degree that the GenAI systems built around these models can match and reproduce conventional examples of this genre. On the one hand, I would argue that this is genre knowledge because the systems were able to recognize and reproduce the

Table 4.2 Rubric Generated by Gemini

Criteria	*Needs Improvement (1 pt)*	*Developing (2 pts)*	*Proficient (3 pts)*	*Exceeds Expectations (4 pts)*
Introduction	Lacks clear introduction of movement and analysis purpose.	Briefly introduces movement and analysis purpose.	Provides clear introduction of movement, analysis purpose, and brief background.	Provides clear and engaging introduction of movement, analysis purpose, and strong background information.
Genre conventions	Fails to identify or poorly explains common features of signs.	Identifies some common features but lacks specific examples or analysis.	Identifies and explains most common features with some relevant examples.	Identifies and thoroughly explains all common features with strong and varied examples.
Purpose and audience	Misinterprets or fails to analyze the purpose and target audience.	Somewhat identifies purpose and audience but lacks depth or clear explanation.	Clearly analyzes the purpose and target audience, explaining their connection to sign features.	Provides insightful analysis of purpose and audience, demonstrating strong understanding of their influence on sign design.
Social context	Fails to connect signs to the social and historical context of the movement.	Makes weak attempts to connect signs to context, lacking clear explanation.	Effectively connects signs to relevant aspects of social and historical context.	Demonstrates deep understanding of social and historical context, providing insightful connections to sign characteristics.
Effectiveness	Lacks evaluation of sign effectiveness or provides irrelevant arguments.	Attempts to evaluate effectiveness but lacks clear reasoning or supporting evidence.	Provides a well-reasoned evaluation of sign effectiveness based on identified elements.	Offers a nuanced and insightful evaluation of sign effectiveness, demonstrating critical thinking about the signs' impact.
Writing and clarity	Writing is unclear, disorganized, or riddled with grammatical errors.	Writing is somewhat clear and organized but may contain some errors.	Writing is clear, organized, and generally free of errors.	Writing is clear, concise, well-organized, and free of errors, demonstrating strong communication skills.
Use of evidence	Lacks or misuses evidence from the signs.	Uses some evidence from the signs but lacks proper integration or explanation.	Integrates relevant evidence from the signs to support analysis.	Effectively and creatively uses evidence from the signs to enhance analysis and demonstrate understanding.

appropriate texts for the situation. This is interesting because I did little to engineer my prompt, so with little information about the rhetorical situation, these systems were able to identify and generate conventional responses. Not only were both systems able to describe what I had to do in response to the rhetorical situation in which a teacher wants to persuade their students to write genre analyses, but they were able to produce writing prompts without me asking them to choose this particular genre. I prompted the systems to create the "thing" I would give to students. Both ChatGPT and Gemini correctly produced a writing prompt suggesting that in the context of my prompts, these systems contain genre knowledge sufficient to identify the writing prompt as an appropriate genre.

On the other hand, the GenAI system works with genre knowledge differently than a human does. Human knowledge is a network of interconnected pathways which represent the sticky relationships between processed observations within an individual's experience, and we learn by encountering new information which "sticks" to the existing network of knowledge that we have learned in the past (Ambrose et al., 2010; Barbey, 2018; Beach, 1999). The consensus among researchers in knowledge transfer emphasizes the importance of cognitive processes like metacognition and mnemonic techniques like the Method of Loci illustrate the ways in which material to be learned is integrated into some pre-existing knowledge structure (Brent, 2011; Dyke Ford, 2004; McCabe, 2015; Negretti, 2012; Reiff & Bawarshi, 2011). Recall then occurs by new experience triggering or not triggering these connections.

Research in "uptake" has illustrated the ways in which new experience might or might not activate past genre knowledge (Artemeva & Fox, 2010; Dryer, 2016; Freadman, 2002; Medina, 2018; Nowacek, 2011; Reiff & Bawarshi, 2011; Rounsaville, 2012; Tachino, 2012). In this context, genre knowledge exists as the representation of lived experience within the network of an individual's knowledge. The "recognition" that Bazerman (2003) discusses is actually the phenomenon by which regions of the knowledge network are activated based on their similarity and difference to the current situation in which one finds oneself. In other words, if I am teaching a technical writing class and my overall goal is to have students learn how to do genre analysis, I might not necessarily think about essays because in this context I want students to compose professional genres and the genre analysis is a strategy to help them figure out these genres. As a result, I will not necessarily select genre analysis essay writing prompts for my students, and I may not even consider them as a potential genre, but instead I will create a different genre entirely.

GenAI "knowledge" is in fact mathematical models of something. As I described in the previous chapter, they are mostly algebraic formulas in which input is processed through a number of mathematical operations yielding an output. The pre-processing and post-processing layers in the GenAI pipeline translate the human language into mathematical values and back into human language. I suggest perhaps a reasonable way to describe the genre knowledge that seems to be present in LLMs is genre conventions and even genre knowledge has been encoded in the LLM. So ChatGPT likely does not have genre knowledge in any human sense. It has encoded genre knowledge to the extent that the texts prompting the genre and the text produced by an individual doing the genre were part of the training corpus. This means that if I want to know how people represented by the training data responded to a particular rhetorical situation, I will be able to get a great deal of information about the genre or genres that respond to that situation from a GenAI system. By interacting with ChatGPT or Gemini, I can learn a great deal about how we have taught genre analysis and how we have prompted students to write genre analysis essays. However, that information does not have the nuance of understanding and intuition that would be available to an individual with genre knowledge. For example, in my experience, students tend to struggle to focus in specifically enough when they are analyzing genres to a particular action or they do literary analysis of genre rather than rhetorical analysis, and so my course content and instructions in prompts attempt to address this particular challenge.

The encoded genre knowledge in Google's and OpenAI's models seems to be somewhat different as well due to the difference of the responses between the two systems. Both provide very conventional explanations of how to assign genre analysis and can produce writing prompts, but they differ considerably. ChatGPT's responses were much longer and less specific to my question. ChatGPT told me not only how to assign a genre analysis task, but also how to teach it. I did not strictly address this, but certainly the prompt and the lesson plan go together. Gemini, however, responded more closely to my original prompt and didn't stray from that particular question very far. Likely this has to do with the different architectures. ChatGPT will have extracted less information from my question because at the time of writing it is a decoder only architecture. Gemini, however, extracted more information and therefore was able to respond more specifically, which is likely due to the encoder-decoder architecture of that system. It is beyond the scope of this project to do extended comparisons, but that would certainly be a useful project, particularly if it is tied to the architecture of each system that is reviewed.

In addition to knowledge structures that help individuals recognize rhetorical situations and identify an appropriate response, genres also play a role in constituting identity and positionality in a particular social context (Bawarshi, 2003; Dryer, 2008; Rounsaville, 2014). GenAI clearly does not have an identity or a position as an individual in a social context. It is a system of computational operations that processes input and calculates output. There are a few important pieces to discuss associated with this aspect of RGS. As Devitt (2004) suggests, when we perform a genre we are negotiating between conventions of our social setting and our rhetorical aims. As we perform this negotiation, we engage our rhetorical awareness and genre knowledge, and also, we involve our own histories, intuitions, body, and sense of self. The performance of the genre gives us a chance to be seen in the social reality insofar as our communicative action makes us visible. What is important is that the identity that we inhabit and the position we occupy is not merely the performance of genre, nor is it merely some display of our self, but rather it is a combination of genre and self as they are constituted together in the moment of the performance.

In *Writing Genres,* Devitt (2004) provides a particularly compelling image to describe the identity and positionality of genre, when she suggests:

> that genre be seen not as a response to a recurring situation but as a nexus between an individual's actions and a socially defined context...genre exists through people's individual rhetorical actions at the nexus of the context of situation, culture, and genres.
>
> (p. 31)

The image of a nexus illustrates an active convergence between what I see as the rhetorical forces that exist within a rhetorical situation. Consider again the example of the writing prompt for genre analysis. I find myself in a rhetorical situation that requires that I ask students to do some work and learn something about RGS. In that situation, numerous forces converge, including the force of my memory of past rhetorical actions—both the ones I've performed and the ones that I've received—my knowledge from my experience teaching and writing, the formal training I received in graduate school, my knowledge of Writing Studies theory and pedagogy, as well as the institution, the institution's memory, and my perception of the students. At some moment, I act bringing these forces together and when I do so, I take on the position and identity of a person who prompts students to write through writing prompts. These forces converge in a nexus that gives me shape and takes the shape of a particular genre that I have selected. GenAI does not do this, but if I use GenAI in my process, then it and the LLM it contains become additional forces that converge within the genre nexus.

The final major trend in RGS regarding the place of genre views genres as active things, or what Anne Freadman (2012) calls "shots" as in tennis shots. Likely genre performs all of these roles. Consider the cover letter I described above. First, for writers familiar with cover letters, the genre itself helps them identify and interpret a particular cover letter when it is presented to them, and on the other side produce a cover letter. Genre provides recognizable conventions that constrain the production of the rhetoric. Genres are also active and do work in a particular social context. For instance, if I ask students to write a cover letter in my technical writing class and they are writing it for me rather than for a hiring manager at a company where they are applying, the genre becomes what Elizabeth Wardle (2009) calls a "Mutt Genre." The cover letter loses the essential elements of the audience and the exigency. Instead of the exigency being a desire to get a job, the exigency is the student's desire to get a good grade, and so their performance of the cover letter genre will be influenced by the wrong thing. Likewise, their audience is different and has different interests than a hiring manager would, so students understanding of the rhetorical situation would not match the genre either. I argue that we might think of genres as currents within discourse. They are recognizable patterns of rhetorical action within a particular social context that allow individuals to persuade and otherwise exert force on others. This notion of currents fits with Sid Dobrin's (2011) fluid dynamics metaphor of rhetoric. The conventions or currents are carried within the memory of individuals and their prior genre knowledge and within the stabilizing material artifacts (like "Patient Medical History Form[s]" (Bawarshi & Reiff, 2010, p. 193)) that we produce to constrain rhetorical actions. When we emphasize the rhetorical action itself that is constrained, then the relationship between GenAI and genre becomes more complicated.

It is clear that GenAI systems can reproduce texts that conform to the conventions of many common genres that were part of the training data. It is also clear that purpose and intention are not factors in how GenAI produces text.

However, we might have a conversation with Gemini to see its ability to engage in rhetorical action. For this I will contextualize the conversation by asking the system to respond as if it were a debate partner, which will it to construct text from the patterns in the LLM that contradict what I write at it. I start by prompting Gemini to "act as a debate opponent in the conversation to follow" to which it replied, "I can tailor my responses to be a strong opponent, presenting counterarguments and challenging your points of view with evidence" (Google, 2024a). The following interactions, I present arguments against grades, and Gemini's responses generally fit into the genre more akin to a teacher giving feedback to a student learning how to debate rather than debate opponent. In the first response, Gemini generates the text, "That's a compelling argument for abolishing grades. Here are some counter-arguments to consider." It follows this with a list of counterarguments, and concludes, "Let's delve deeper. How would you address the need for clear communication of student progress without grades?" In this exchange, I don't see a debate partner who is making counter arguments directly against what I've said, as I might expect in this particular genre system. Instead, it takes the stance of a teacher or tutor who is helping me formulate my argument and account for potential limitations. I do believe that this is a good response, and I would find this a useful tool if I were trying to create a strong argument, but since I explicitly identified the rhetorical situation to be one where Gemini would generate text as if it were a debate partner, the system does not satisfy particularly well that rhetorical exigency. I suspect that a large part of why this is the case is that Gemini's models probably have been trained and biased toward friendliness and against arguing against the user.

The point of this exercise, however, is to illustrate that while LLMs have effectively encoded genre knowledge and can provide exceptional insight into

https://g.co/gemini/share/e44241a8f708

Figure 4.3 Scan to access Gemini Conversation: "Ready to Debate".

language that has already been used, GenAI is not performing a particular rhetorical action because it is not occupying the correct genre in its response. This, of course, makes sense. The amount of information encoded in LLMs is large enough to contain some genre knowledge, which is interesting because it suggests that the genre conventions are not purely interpretive lenses, but rather appear naturally in the text. Further research into the genre knowledge encoded in LLMs would be needed, however, to see to what extent these conventions are translated from the "psychosocial recognition phenomena" to the textual features that were extracted during training.

4.3 GenAI in the Network

In addition to considering how GenAI systems are able to model language and provide insight into genre conventions, we might also consider the way these systems influence the way that individuals can produce meaning as they use them. Several theoretical foci help us consider the way that complex systems are involved in producing rhetoric. One way that we might approach this is focusing on the interactions that occur between people, spaces, technologies, and discourses when rhetoric is occurring. Cress and Kimmerle's discussion of GenAI focuses on the notion of "knowledge-building" and focus on the "*co-evolution model*" (2023, p. 610). In this argument, they suggest that making knowledge is not the result of a single individual's cognition, but rather the interaction between the individual cognition and the society in which the knowledge-making is occurring. In the model of "co-constructing" with GenAI, the idea is that while perhaps GenAI is not doing any meaning-making, it is does provide some foil against which an individual's cognition might work. The example above in which I tried to have a debate about abolition of grades with Gemini illustrates this fairly well. While Gemini was unable to respond to the rhetorical situation I presented it by engaging in a real debate with me, it did generate a number of counter arguments that did honestly inspire me to think more and further complicate the way I understand grades. This is possible because GenAI can effectively reproduce the language it was trained on, and the arguments contained therein. As long as I am making meaning about some topic that already has significant literature surrounding it, GenAI will be able to fairly accurately reveal the various things I have said and potential answers to blind spots that I might have in my argument. Interactions within a system have been the focus of some research in Writing Studies. Karen Barad (2007) extends quantum mechanics into the realm of social theory suggesting that the "world in becoming" or more simply the world we live in is the product of "complex agential intra-actions of multiple material-discursive practices or bodily production" (p. 140). Barad's ontology emphasizes the way that the materiality of the space, the technological apparatuses, and the systems of discourse interaction to bring the world—at least as we know it—into being. I suggest elsewhere (Medina, 2017, 2018) that this extends to the way we make meaning as well. For a simple example, the technology I am using to write is not merely a medium or conduit into which I make meaning, but it is an active participant in the production and has an effect on the meaning that is produced.

This notion of writing as the product of complex system exists in conversation with work already being done in Writing Studies on network theory (Dingo, 2012; Latour, 2007; Rice, 2012; Spinuzzi, 2003, 2008), writing ecologies (Dryer & Peckham, 2014; Edbauer, 2005; Fleckenstein et al., 2008), and writing as fluid dynamics (Dobrin, 2011). Rhetorical theory that focuses on the network reveals the ways that rhetorics and genres operate within and circulate. Dingo (2012) analyzes the ways in which feminist rhetorics are circulated across global networks focusing on the ways in which these rhetorics come into contact with local customs. Through this analysis, she reveals the ways they might both impose values and are distorted by the contact. The movement of feminist rhetorics across global social networks is interesting because it illustrates the way in which these ideas travel, and if we apply this approach to the language represented by GenAI, there is a particularly relevant potential for the language model to have significant impacts as it is used across social networks. While the LLM is representative of a massive collection of text, this corpus itself is representative of only one subset of language. Dobrin describes this as "exclusionary bias" suggesting that the content that is excluded from the LLM produces significant bias in the way that language is represented (2023, p. 122). I would suggest that there is also inclusionary bias, meaning there is both a decision-making rubric that prevents texts from being included in the corpus, but also there is another rubric that increases the likelihood that some texts will be included. For instance, Wikipedia is frequently included in natural language corpora because it is easy to scrape, the content is high quality due to the review process involved in contributions to Wikipedia, and the licensing of the content minimizes legal concerns. At the same time, Wikipedia's contributions guidelines introduce a linguistic bias when this is a major source of content for language model training. Likewise, it is possible and common for those building LLMs to include particular texts to introduce bias intentionally. The concern with the networked rhetorics that Dingo discusses pertains to GenAI in the sense that as texts generated by GenAI proliferate across discourses and publics, these texts will spread the ideology or the discourse that is encoded in the model. Likewise, as I suggested in the previous section, LLMs seem to contain genre knowledge, so following Clay Spinuzzi's (2003) discussion of tracing genres across networks, it is likely that GenAI is also transmitting this genre knowledge across the social networks of the individuals who use it. Using Dingo's and Spinuzzi's approaches to tracing rhetorics and genres across social networks, researchers could more clearly see the effects that GenAI is having on discourses.

Research in writing ecologies and complexity theory offer us a way to think about how GenAI operates within social networks and how its writing exerts force on the audiences therein. These theories deploy a metaphor of writing that positions the writer as an enmeshed member within an ecosystem, meaning the writing exists as the word and output of relationships between "writers, texts, and environments" (Fleckenstein et al., 2008, p. 393). This means that even though I emphasized the importance of the agency of the writer in the previous section, the writer does not exist in isolation but is an integrated member with a complex apparatus that includes the person; the material environment; the discourse, the knowledge, the history, and the people around them; and the technologies they use. The writer's agency exists within this ecosystem. In a similar vein, Sid Dobrin extends complexity theory "[offering] a perception of writing as a complex 'liquid' system" (Dobrin, 2011, p. 179). This notion is compelling because it emphasizes the way the potential for movement, flow, and the exertion of force by language. Since people tend to find themselves in rhetorical situations—rather than actively seeking them out—we might think of these situations emerging from the flow of everyday life. Likewise, the writer's response acts as a counter force that responds to the demands of the situation. This notion is compelling because it suggests that the rhetorical action that responds to the situation, or the persuasive act that attempts to convince the audience to do or believe a particular point of view—might be seen as a force or a current that moves within the constraints of the various other forces that are acting upon the situation. One way to look at GenAI is as another tool within these ecosystems. If a writer uses GenAI, for example, to help them craft arguments, the technology will give them greater insight into the discourse and history surrounding that argument. In the interaction I described above over grading, Gemini was able to clearly illustrate the various arguments that might be made and therefore the potential perspectives that might exist within the ecosystem as I write my argument. As long as I do not rely too heavily on the language generated by GenAI, I will be in a position to navigate the rhetorical ecosystem quite effectively.

The major concern that will require further research is over the effects of GenAI on the rhetorical ecosystems where it exists. Consider writing without GenAI. A writer has their knowledge and experience, which is the culmination of a history of interactions with content and past rhetorical situations and so on. Writing Studies, particularly the research on Transfer that focuses on pedagogical memory illustrates the importance of this past experience (Jarratt et al., 2009). Within the rhetorical ecosystem the writer is also interacting with potential peers who might be reviewing their writing. Each of these peers will have unique histories. These histories produce perspectives that are limited to whatever information these individuals have read and whatever situations they have experienced. These histories are also limited to the uptake of the individuals and what parts of these experiences and knowledge they have assimilated (Ambrose et al., 2010; Fiscus, 2017; Freadman, 2002; Macklin, 2019; Reiff & Bawarshi, 2011; Rounsaville, 2012; Tachino, 2012). This means that when an individual encounters rhetorical situations they filter the information they gather about those situations through their experience with antecedent genres. Additionally, their perspective is limited by the way that knowledge has been assimilated, which is a complex process involving connecting new information with the network of knowledge that the individual already knows. This is all to say that the human elements of the rhetorical ecosystem contribute to the writer's ability to respond to a rhetorical situation through these various filters and knowledge structures. Additionally, the ecosystem includes the institutions and discourses, which are made of people, but also institutional memory. These structures constrain and guide the way that rhetorical forces can be exerted. For instance, an institution might have specific rules governing what sorts of speech can be made, thereby rightly limiting the writer. Finally, the technologies used to transmit writing play a significant role in the way rhetorical force can be exerted. Consider internet communication platforms like social media and discussion boards. These have expanded the reach that any individual might have otherwise had. All of these taken together form the rhetorical ecosystem and it is through their interoperation at focused on the site of the writer that the writer might respond to a rhetorical situation and exert their rhetorical force back into the ecosystem.

Something that will require further research is the effect that GenAI will have when introduced into the rhetorical ecosystems. While LLMs treat language in a different way than humans do, they do encode information and model a perspective on language that is much broader than might have been possible in the rhetorical ecosystem previously. This means that the advent of GenAI introduces into the writing landscape a system that can effectively generate language based on a broad representation of natural language. Prior to the wide availability of GenAI, if I were to write some argument like the one about grade abolition that I describe above, my argument would be limited to what I know and what my peers or reviewers know. At the same time, our perspectives will be diverse and unique, and we will be able to perhaps generate a complex and novel understanding of the topic. When I use GenAI to support my writing, I can immediately gain access to a fairly extensive summary of all of the existing perspectives of an argument—unless not very much was written about it in the dataset on which the LLM was trained. The main limit placed on GenAI is the number of tokens the system will create—this is a configurable limit. This means that GenAI provides extremely easy access to thorough summaries that represent the conventions of the data it was trained on. Additionally, and perhaps more significant, the language generated by GenAI is conventional. The question I am approaching is, if GenAI is so effective at responding to questions, even if its output is overly conventional and is incapable of producing anything new, what sort of effect will it have on the rhetorical ecosystem? Will it carry an oversized weight and perhaps dominate the production of writing? With responsible use, I think there is little worry of this, but if GenAI is used as a shortcut thereby disrupting the writer's contribution, there may be a reason for concern. As GenAI becomes more extensively integrated into our rhetorical ecosystems, it will be an interesting avenue of research to see what affect that technology has on the ecosystem, particularly in its potential to homogenize what and how things can be said.

Note

1 Hallucination is probably an inappropriate term because it implies the object of the hallucination is not there. When GenAI "hallucinates" the object is there insofar as there was a pattern of information in the model that did generate the erroneous output.

5 Risks and Opportunities in Pedagogy and Research

5.1 Intellectual Property

Much of the scholarship in Writing Studies on GenAI focuses on the pedagogical implications answering the question, how should we approach GenAI in our classrooms. The main topics that have been covered in the pedagogical implications include concerns over academic integrity and plagiarism, and how might we teach students to use GenAI effectively. In this chapter, I will discuss these topics and conclude with potential paths for future research.

Nearly all scholarly articles and policy statements that I have encountered on GenAI contain, and even foreground, a concern over academic integrity and plagiarism. I suspect that the prevalence of the issue of plagiarism in discourse surrounding GenAI stems from the importance placed on intellectual property (IP) in institutions that are inevitably immersed in capitalist notions of property (Jessop, 2018; McKenna, 2022). Alternative approaches, like those proposed in Creative Commons licenses, run outside of or in contrast to capitalism and are probably preferable for digital and multimodal composition that is done with or without GenAI (Johnson-Eilola & Selber, 2007, p. 399). I will not speak to the ethical implications of IP ownership, nor will I speak to capitalism, in general, but those are discussions worth having. Plagiarism means not only taking IP, but also cheating the system of assessment through writing that is so critical to rank, privilege, and gatekeeping apparatuses in academia. If GenAI is a new form of plagiarism, then it poses a threat to the very systems of assessment that are core to our institutions—particularly in the Humanities and Social Sciences where writing-based assessments are the norm. It would be useful to review the many ways that university statements on GenAI position the technology in relation to IP and plagiarism. I want to set aside, however, concerns of plagiarism related to academic dishonesty because I suggest that, like many others, we address this as an opportunity for learning and engagement rather than a moral failing that needs to be policed (Evans-Tokaryk, 2014; Jamieson & Howard, 2019; Nelms, 2015).

DOI: 10.4324/9781003493563-5

Instead, I want to discuss here the more interesting component of the relationship between LLMs and the IP they were trained on. Jean Liddell (2003) defines it as

> the act of using someone else's words, ideas, organization, drawings, designs, illustrations, statistical data, computer programs, inventions or any creative work as if it were new and original to you; this includes real and intellectual property and public domain material.
>
> (p. 49)

Liddell continues to suggest that plagiarism can range from intentionally submitting writing that is not one's own without attribution to incorrect use of quotations and paraphrasing. In this definition, honesty and cheating are present, but more interesting are notions of ownership and property. According to the World International Property Organization, "Intellectual property (IP) refers to creations of the mind, such as inventions; literary and artistic works; designs; and symbols, names and images used in commerce" (*What Is Intellectual Property (IP)?*, n.d.). In other words, intellectual property is something that was created by the mind of an individual or group of individuals, who are by default the owners of that thing—unless some other arrangement was made. The World Trade Organization defines intellectual property rights as "the rights given to persons over the creations of their minds. They usually give the creator an exclusive right over the use of his/her creation for a certain period of time" (*WTO | Intellectual Property (TRIPS)—What Are Intellectual Property Rights?*, n.d.). These rights are protected by things like copyrights, patents, and trademarks. This is all to say that intellectual property is a notion that when a creator creates a thing, by default, they exclusively possess the rights to that thing. In the context of plagiarism, the notion of intellectual property suggests the issue at stake is theft.

The question we might ask then is NLG theft of one's intellectual property. If language models contained the original text itself, it would be pretty straightforward because the system would be literally taking text from the source and presenting it as its own without citation. Instead, the language models used in GenAI were trained on massive quantities of natural language, mostly appearing in the somewhat open forum of the internet. GPT 3 was trained on 45TB of data from the Common Crawl (Brown et al., 2020). More recent crawls brought this into the range of petabytes. It would be unfeasible to run inference using an LLM if this training data were actually present in its original form in the model. Instead, this training data was cleaned, processed, and then run through the transformer architecture yielding language model derivatives. This means that even if an LLM ingested a particular sentence from a Wikipedia article, it would be impossible for me to locate that sentence in the model because it does not literally exist in the model. Instead, the encoded values of that sentence were used to adjust the weights of the language model. In other words, it would be better to see the language models as mathematical derivatives of natural language, which is literally what they are. In other words, the model is mathematical values that were adjusted automatically based on the mathematical encoding of the training set.

Since LLMs do not directly contain the intellectual property of anyone—except perhaps the researchers who developed the algorithms—but instead are derived through the processing of intellectual property that was accessible on the internet to web scrapers, the discussion of plagiarism as an intellectual property concern would be better addressed as one of fair use and the creation of derivatives. The first question worth reflecting upon is the question of accessibility and more largely the ethics of web scraping. Common Crawl data and other large corpora are often the product of web crawlers or automated applications that iteratively visit websites on the public facing internet and copy the text content from those sites. Building such a bot does not require much technical expertise, and the human hours involved in collecting the data are insignificant. In Python, developers have options like Scrapy (Scrapy Developers, 2024) which can quite easily walk through a site or number of sites extracting textual data. However, one would not need to do that for a base dataset because projects like Common Crawl regularly traverse the internet and provide that data openly and freely. In other words, when publishing content to the open web, one probably should be aware that this content might not be read only by humans but might also be accessed by web crawlers. This awareness is a topic worth discussing and could be addressed as something we discuss in writing classes. It could also be addressed in the software we use to publish content. While many technologies where we publish content require that we sign privacy policies, these are often inaccessible. Google's Gemini perhaps has the right approach where they place a warning that our input text will be used to further train their models and will be read by human researchers in a very visible position on the screen.

If we are aware of how our content is accessed, the writer does still have some control over it. Websites can include a robot.txt file that web scraping bots are supposed to read that can provide directives regarding the behavior of those bots. Common Crawl provides a guide on how to prevent one's site from being crawled by simply adding a rule in the site's robot.txt file—the file that defines how bot users should interact with one's site. Less "friendly" web crawlers might ignore the robots.txt file, and in that case one could add anti-bot technologies like ReCAPTCHA, registration walls that require email confirmation, and paywalls to make it more difficult for bots to access the site. In other words, if I write and publish something to the open internet, it is possible for me to prevent that writing from finding its way into these massive datasets that could be used to train LLMs. I could add rules to my robots.txt file, I could add any number of anti-bot technology, or I could simply put my site behind either a paywall or registration wall requiring users to register and possibly pay to gain access. Certainly, it is possible to overcome these obstacles, but if my main concern is privacy, then putting information that I want to keep private into a digital format anywhere on the internet is risky. The above-mentioned countermeasures are sufficient to keep my content visible to humans and private from bots.

One of the concerns that we should have as writers and that we should teach our students about in our writing classes is the notion of intellectual property and privacy in digital composition. How writing is produced and knowledge is disseminated is a significant part of technical communication (Doheny-Farina, 1992; Gaines, 1993; Hannah & Lam, 2016; Spinuzzi, 2008). In a world where a vast amount of human knowledge and human communication are done through digital modes and cross networks; engagement with these networks is crucial. Multimodal composition interests itself in the communication channels and the effect those channels contribute to the overall genre enacted by the rhetor (Shipka, 2011), I suggest that we should also be concerned with who might be listening to that rhetoric and how it might be accessed into perpetuity. In the case of web crawlers, writers composing in digitally accessible spaces should be aware that their writing might be retrieved by non-human processes and used for various unintended purposes ranging from language research to advertising. While regulations are certainly an important part of this issue, and while scholars in fields of Writing, Communication, and Composition should be involved in these regulatory decisions, we also must be prepared as writers and prepare our students with some level of understanding so that they might gain greater agency over the dissemination and use of their writing. In the next chapter, I will discuss what this might look like when I make recommendations for pedagogy and research. I also suggest that concerns of privacy that are rooted in what data is retrieved and how that data is disseminated in the creation and use of GenAI is a potentially valuable topic for further research.

The second key component of intellectual property beyond just how it is accessed is what is done with that property by the audience. Clearly intellectual property that is shared publicly implies that those who can access that publicly shared information have the right to consume that information. If I publish an article on the internet, I am giving anyone online free access to read that article. Beyond accessing and reading intellectual property, other uses might be allowed without consent from the creator. These uses are largely covered by Fair Use frameworks. These consider things like the individual or entity using the IP from another creator, the nature of the use, the IP itself, and the consequences to the creator. If I write a series of lesson plans for a writing class, and I share those lesson plans, fair use might look very differently upon a teacher borrowing my lessons and reviewing them or using them in their own teaching and a company that compiles my lesson plans into a book for sale. Creators have addressed this by attaching copyrights or licenses to their creations. Many of the software frameworks and LLMs are in fact released as "open source," which means they are released to the public with licenses that govern the use of the software. These licenses might specify that the software is released as is and the creator is not liable for any potential issues, it might specify whether the software can be modified, close sourced, used for commercial projects, and so on. Similar work is done by artists and writers by attaching various copyright or copyleft statements to their work. Both open-source licenses and copyright and copyleft statements provide language that allows for derivative works.

So, we cannot exactly think of LLMs as a form of plagiarism insofar as they are copied or even patchwritten versions of the original text. Instead, we are left with a beginning of how we might consider the relationship between LLMs and language in the frame of plagiarism and IP: derivation. These models are literally derived from calculations done on training data. This is useful because it allows us to abandon notions that LLMs and GenAI is selecting the next most probable word from the training data as if the model contains the training data. It is not a system that selects from an existing corpus, but rather generates based on patterns extracted from the corpus. With this understanding, we are in a better place to discuss the rhetoric of GenAI. I suggest that we might consider LLMs derivatives of natural language and the generated texts they produce as descendant derivatives. Descendant derivative refers to the various layers that might be included within the NLG pipeline. Consider a parameter-efficient fine-tuned pipeline that includes an encoder, a decoder, and a head layer. The encoder and decoders used in this architecture are general purpose and are derived from general language—some massive dataset that was used in their training. The parameter-efficient fine-tuning adds another one or more derivatives of domain specific data. So, the resulting text generated by this pipeline will be derived from calculations done on input data through several steps, each step contributing to the output text. I suggest that a major avenue for future research, particularly for scholars interested in IP, is this derivative relationship between the training data and the output.

5.2 GenAI as a Threat to Translingualism

Academic Integrity and Intellectual property are two concerns that appear regularly in the discussion of GenAI. While these are significant risks affiliated with the technology, a more significant one is the challenge to the antiracist, anticolonial struggle against the dominant role of what Suresh Canagarajah calls "Metropolitan English (ME)" in our writing classrooms and institutions (2006). I recognize that while Translingualism and antiracism are central to my own teaching practices, these topics have not been central to my own research work. However, I believe that the threat to students' linguistic rights and identities is so significant that I want to address it briefly here with the hope that it is further explored by scholars with greater expertise in Translingualism will take up this topic.

The role of GenAI for multilingual students, particularly students with limited experience with the privileged English used by academic institutions and their representatives is a significant issue. Institutions in English speaking areas frequently use language assessments—both in the form of the TOEFL test and written admission materials—as gatekeeping technologies. According to ETS, the TOEFL test has been taken by more than 40 million individuals for entrance into more than 13,000 institutions around the world (*TOEFL Test Takers*, n.d.). While GenAI cannot help with the TOEFL test, it could help students conform to the conventions of ME in their writing tasks. Some see this as a boon to such students. When discussing the potential problems with simply banning GenAI, Northwestern University suggests, "students for whom English is not their first language [who] may rely on GAI to enhance their grammar and fluency in academic communications" (*Use of Generative Artificial Intelligence in Courses*, n.d.). GenAI can help students overcome the obstacle of privileged English in the institution, but it is solving the wrong problem. For several decades, translingual perspectives have challenged the place of ME because doing so in institutions establishes hierarchies in highly problematic ways. Suhanthie Motha suggests that "it could be argued that nowhere within school walls do empire and racism find more vociferous expression than in the location of ESOL" (2014, p. 48). Promoting a ME in these classes and across the institution does the work of erasing other languages that students might use, and beyond this it incorporates the same kinds of heteropatriarchal, colonial, and white-supremacist structures upon which the discourse was founded (Smith, 2016). The antiracist, anticolonial work of Translingualism encourages teachers and institutions instead to take seriously students' right to their own linguistic practices (CCCC, 2018). This appears in several ways. Nancy Bou Ayash (2020) addresses the potential use of translations in "paratexts" as "critical sites for interrogating and intervening with the dominant monolingual discourses and norms that dictate the production and reception of academic written work." In other words, instead of encouraging linguistic homogony through translation, Bou Ayash suggests language difference can be an opportunity to make meaning in ways that challenge the dominant language practices of the discourse. In addition to disrupting linguistic hegemony, translingual scholars suggest that translingual pedagogy and increasing diversity in institutions "will drive instructors to seek flexible pedagogical approaches suitable for shifting and oftentimes unstable contexts" that will them "the ability to explore connections and points of compatibility between theories and practices, communicative needs and complex literacy backgrounds" (Sánchez Martín et al., 2019, p. 154). Again, translingual perspectives highlight the value of linguistic diversity and the opportunities that it presents for creating novel meanings. This diversity and these opportunities are threatened by GenAI.

GenAI has the potential to resolve the challenge that many students face in having to work within discourses dominated by ME. Students need only prompt one of the various GenAI platforms for a piece of writing that follows the students' desired line of argument, and it will generate a text that conforms to the conventions of ME because the LLM was trained on this English. However, this reinforces the colonial project of English education, silences non-dominant voices, and disrupts the opportunities for new ways of making meaning—and the meanings that could be made through translingual writing. In the context of writing education, the use of GenAI should be approached carefully to counteract this risk. Writing studies need significantly more research into the effects that GenAI has, particularly on underrepresented or multilingual students, and development into ways to position GenAI as a useful tool without disrupting the opportunity of translingual writing.

5.3 Class Is in Session

So far in this chapter, I have highlighted some of the risks associated with the use of GenAI in writing classes. In spite of these risks, GenAI is already being used widely by students and professionals. In educational contexts, one survey conducted in from May 2024 found that almost 50% of teachers, students, and parents are using ChatGPT at least once a week (*The Value of AI in Today's Classrooms*, 2024). It seems fairly clear that GenAI is only going to become more of the pedagogical landscape, and this gives educators the opportunity to position it in a way that best mitigates the risk while gaining the benefits. While I have identified some of the limitations in the scholarship in Writing Studies on GenAI, one strength is the pedagogical recommendations. In particular, Cress and Kimmerle's (2023) focus on co-construction of knowledge through conversations with GenAI is an interesting path. Since the LLMs in GenAI model language fairly effectively, there is a place for interacting with one of these systems as we generate our line of arguments. The challenge here, as I discuss above, is to ensure that the content created by the GenAI system does not overpower the writer's ideas. Perhaps the most thorough discussion that I have encountered of the uses of GenAI in the writing classroom is in Sid Dobrin's book *AI and Writing*. In the fourth chapter, Dobrin outlines ways in which GenAI might be used in the various phases of the writing process. During the "prewriting" phase, writers might use GenAI to help them find the beginning of their writing (Dobrin, 2023, p. 64). During "invention," writers explore a topic and find a place to stand therein. Each movement in Writing Studies has identified a particular approach to "prewriting," and GenAI can certainly help with this. LLMs probably have encoded a considerable amount of information about many topics students might write about. I suspect however, that if the first step in writing is to turn to ChatGPT for "inspiration," there is the risk that the writer might be too inclined to accept the conventional ideas and unwilling to present their own ideas if they are heterodoxic.

Dobrin also suggests that GenAI can play a role in research because it "can direct you to relevant information, identify important sources for further reading, and direct you to pertinent conversations" and "summarize key points for research" as long as one is aware of potential risks of hallucinations (2023, pp. 65–66). One of the major challenges that writers encounter when starting new research is understanding the shape of the field—what has been said and where there may be gaps. As long as the LLM has been trained on data relevant to the field that students will be writing about, it will be able to summarize that information. I suggest that this is perhaps one of the better uses of GenAI for students and researchers working in a new subject area alike. I also see an opportunity for student/teacher interactions to exist during this phase. GenAI might hallucinate if it contains incomplete information about a particular topic, but students may not have sufficient experience and knowledge in the field to both identify hallucinations and ask meaningful questions about the generated content. With teacher engagement, there is a strong opportunity for students, GenAI, and teachers to interact as students gradually find their position from which they will begin their research. Depending on the level of the students, this could begin with the teacher providing students with prompts or students generating their own prompts that they input into one of the GenAI systems. After several turns, students could submit these interactions to the teacher who could provide feedback and guidance based on the interactions. Through this process, teachers will be able to model effective engagement with GenAI and students will learn a considerable amount about the shape of the field. Within research uses of GenAI, it might also be useful for disciplines to train or finetune LLMs on research in their field. For instance, Gemini and ChatGPT both effectively summarize Rhetorical Genre Studies, but further training on scholarship from the field could improve its representation of that subset of knowledge. Perhaps this could be a project carried out by the academic conferences like the CCCCs and MLA. One particular technique that could be useful might be to use Retrieval Augmented Generation, which allows a developer to provide additional context documents to the inference pipeline that is using a more general LLM (Merritt, 2023). At runtime, one could submit a corpus of scholarship to such a system and accuracy and reliability in the responses to prompts about the topic would be significantly enhanced.

Once again, Dobrin's discussion of the use of GenAI for drafting is reasonable. He suggests that while this technology is useful in the drafting process, "[o]utputs [from the GenAI system] should *inform* your drafting; they should not themselves be your drafts" (Dobrin, 2023, p. 67). It is, I believe, critical to discourage students from copying and pasting directly from the output of GenAI. I suggest that in the drafting phase, GenAI poses the greatest threats. In previous chapters and sections, I illustrated the ways that LLMs encode conventional language and reproduce it when prompted. I have also discussed the threats, particularly to students for whom writing has been a significant struggle. The challenge with using GenAI extensively for drafting is that the generated text will likely be more fluent or clear than an early draft that a student writer might produce. This is because GenAI is particularly adept at creating writing that is conventional, correct, and formally at least, fairly well polished. When faced with a choice of the awkwardness of writing that is the product of struggling in interesting ways with challenging ideas and the highly conventional and formally correct but less interesting output of a GenAI system, not choosing the latter would require a fair degree of confidence in one's ability to improve the interesting mess of the former. Beyond this, there is a significant concern over student voice—whether this be "voice" according to expressivists or "voice" according to translingual scholars. A critical component of teaching writing involves providing opportunities for students to find their way of saying things. There is a significant risk of students losing their voice in the face of GenAI during the drafting phase. One example of pushing back against GenAI's drafts that might be useful is an exercise that I have done in my own classes involved having students prompt ChatGPT to write a draft of an essay. I then asked students to provide feedback and criticism to the draft. In this way, I asked students to take up the position of reviewer and teacher rather than novice writer, and in doing so they had the chance to write the genres of marginal and end comments. By taking up a critical stance, students may be less likely to simply take the generated text as good writing. While further research is required to discover how GenAI affects students' ability to find their voice, I feel like this activity does encourage students to approach GenAI from a critical stance, and it allows them to begin by pushing back against the text generated by ChatGPT.

GenAI can also be quite useful during the revision process. Dobrin suggests that "[p]erhaps one of the most beneficial ways GenAI can assist in revision is by providing feedback" (2023, p. 71). I am somewhat biased in this area, but I do agree.[1] However, this is a particular use of GenAI that needs to be clarified. When I give feedback to a student's writing, I understand the writing in a human way. I am aware of the rhetorical situation, and I can understand the argument—rather than simply match it to a pattern that exists within an LLM. Furthermore, I hopefully am familiar with the student, the work they have done in the class previously, and I will probably have a sense of how they will receive feedback. When I write my feedback, then, I am performing a rhetorical action that is attempting to persuade the student to react in a way that will be beneficial to their writing and their learning.

When we ask GenAI to generate feedback the process works in an equivalent way as when we prompt GenAI to write anything. Our prompt, which includes both our particular question and the document that is up for review, will be preprocessed, tokenized, and encoded into a collection of word embeddings. These word embeddings will be passed through one or more layers in a Transformer model, and the output will be decoded back into natural language. At no point does the system "understand" the writing to which it responds, just as it does not have any insight into the context or the student as I might have. All of that said, at least anecdotally, GenAI is able to generate feedback that is conventional but also correct. Understanding as we do how GenAI works, the reason that it can produce valuable feedback is because writing done in education settings is fairly conventional. Many writing assignments follow similar patterns asking students to do similar rhetorical moves. Students' responses to our prompts and the issues that feedback might address are also fairly conventional unless we have a radically open-ended assignment. I suspect that a fair amount of this academic writing alongside feedback was encoded during the training of the LLM, and as a result, the GenAI systems using this LLM will be able to match conventional student writing and generate the corresponding matching conventional feedback. In other words, GenAI provides feedback effectively on any given piece of writing to the extent that the training data contained writing with "matching" issues. While this might seem like a critique, I suggest that this is a valuable use of GenAI. First, students can receive conventional feedback in a matter of minutes. Second, there is an opportunity for teachers to use GenAI in conversation with their students and their students' writing. Students might submit their writing and generate feedback using one of the GenAI systems. This writing and feedback can then become the centerpiece for a conversation on the issues raised in the feedback and also on what to do with the feedback that the student received. Generated feedback is not particularly new, but LLMs have substantially increased its potential. Prior to the advent of LLMs, little research was conducted on the effectiveness of this kind of feedback on student learning (Stevenson & Phakiti, 2014). At this point, as GenAI gains considerable traction, it would be useful to research this further.

While we are just scraping the surface of what is possible with GenAI pedagogy, I want to address one other potential use, and that is rhetorical or genre awareness, which are useful for effective genre performances (Artemeva & Fox, 2010; Reiff & Bawarshi, 2011; Rounsaville, 2012). In the last chapter, I illustrated that LLMs have encoded rhetorical and genre knowledge. A particularly interesting use of GenAI would be to use the system to gain insight into the genres themselves. For instance, if I wanted to write a genre I've never written before, I might prompt a GenAI system to both write a sample and also explain how the sample is working rhetorically. I could then perform rhetorical or genre analysis on the sample text and compare my analysis to the insights provided by the system. This saves me the time of gathering a number of samples of the genre, reading all of them, and identifying patterns and divergences. If the genre is relatively conventional, and as long as enough samples of it are included in the training data, there is a good chance that the sample generated by GenAI will closely reflect the conventions of that genre. From there, I can decide how exactly I want to perform the genre, but that decision will be informed by insight into the conventions. The crux of such an activity would be to help students develop skills in genre analysis and an analytical perspective rooted in genre theory.

I want to conclude this section with a brief discussion of the first course I designed that placed GenAI as a valuable tool. The course was an intermediate rhetoric and composition course focused on digital rhetoric and was taught in an asynchronous, online format at a large research University on the West Coast of the United States. The course lasted four weeks but was accelerated, so students earned a full quarter's worth of credits in those four weeks. Students in the course were either English majors taking an elective in their major of study or students from other disciplines completing required composition credits. During the first module, students were invited to find some rhetorical theory lens that they might use alongside digital rhetoric to frame the writing and thinking they would do throughout the class. Since the course was intermediate level, there was an assumption that all students had encountered a survey of rhetorical theory in the past. After this re-immersion into rhetorical theory, students read Stephen Wolfram's "What is ChatGPT Doing … and Why Does it Work?" (2023). This piece was perhaps somewhat math heavy for students, but having a thorough overview of the technology was useful. At the same time, I ask students to experiment with the GenAI system of their choice by asking it a series of prompts on some topic of choice trying to identify strengths and weaknesses of the system. In the future, I would perhaps assume that students have already used GenAI, and I would focus this assignment on trying to get the system to make mistakes. Once students completed their experimentation, I asked them to write a reflection on their experience referencing Wolfram. The second cycle of the course asked students to identify an argument and use GenAI to augment their writing process. Since the course only lasted four weeks, students wrote a draft and one round of revisions.

The most important component of this class was the reflection component. The goal was to gain a better understanding of how to use GenAI as a tool in the writing process, and so reflection gave students the opportunity for metacognition. This metacognition is a key component of learning, in general, and developing rhetorical knowledge and skills, in particular (Ambrose et al., 2010; Beach, 1999; Negretti, 2012; Nowacek, 2011; Reiff & Bawarshi, 2011). While this course was limited and since human subjects research was not conducted, the kinds of conclusions that I would make from it are equally limited. However, it was clear that GenAI can be explicitly included in writing activities, and when it is done with clear boundaries and a sense of how the technology can be used, GenAI can be a useful tool. However, if the goal is to practice skills like style and mechanics, and especially if the students are less inclined to resist the tidy prose generated by the system, GenAI could become a greater hinderance. As policymakers and educators make choices about GenAI in their pedagogies, it is important to not simply allow or ban it, but consider how it can be put to use.

5.4 Research Implications

While much of the scholarship in Writing Studies on GenAI has emphasized the practical, pedagogical concerns surrounding GenAI, there is a great need to conduct more research on the technology and its effects on the classroom. There are numerous lines of research on GenAI that could be tremendously fruitful. My goal throughout this book has been to provide a more technically grounded and theorized foundation for work on GenAI in Writing Studies. However, I, alongside the other research that has been published, have barely scratched the surface. Perhaps the most approachable projects would use the UI of GenAI chatbot to gain insights into the systems and the LLMs they contain. For instance, in the section above, I illustrated the way that Gemini and ChatGPT respond in different ways to my request to generate a writing assignment. This kind of experimentation could be done with various assignments across the disciplines, and it would be particularly interesting to compare the output from conventional assignments and less conventional assignments. Since the LLMs do represent dominant natural language, performing discourse analysis or critical discourse analysis could yield useful insights into the discourse represented by the dataset that was used to train the models.

In addition to insights into discourse, work has already been done with language models to better understand language itself. In fact, much of the research work from Computational Linguistics that is behind the NLP tools that have been built and the ways language is modeled stems from an interest in gaining a more generalizable understanding of language itself. For instance, one of the implications of the possibility of LLMs and the fact that they do seem to represent language in some critical ways support structuralist perspective on language (Gastaldi, 2021). LLMs are the product of automatically identifying patterns that appear in human language, which suggests that there are naturally occurring structures in language that produce those patterns. Likewise, since some genre and rhetorical knowledge have been encoded in the model, there is potential to investigate the way these rhetorical features exist in the models. One of the main obstacles to researching the way that LLMs represent language is their vast size and the hidden process by which they were trained. Further research into the LLMs could be conducted directly with the software itself. This approach would have a somewhat higher barrier of entry than using the GenAI UIs to gather data, the main one being learning about the models, the frameworks that can be used with the models, and the programming languages that are prevalent—specifically Python. Fortunately, there are numerous resources available to learn these.

In addition to researching the language and rhetoric of GenAI and LLMs themselves, there is a significant research potential for understanding the phenomenology of GenAI. It is clear that people are using the technology, and it would be worthwhile to better understand how they experience it and what it does to discourse. For instance, I suggest that there might be a significant risk to student agency when they rely heavily on GenAI or even think of GenAI as a collaborator. Understanding how students negotiate their interactions with it could help answer questions about the way teachers should position GenAI in their classrooms. One possible obstacle is students' willingness to admit their use of GenAI, particularly when nearly all institutions are considering uncited uses of it as plagiarism. This research into the phenomenology of GenAI in students could occur at each of the stages in the writing process that Dobrin outlines. Beyond students, it would be useful to investigate how professional writers work with GenAI and how the public sees GenAI. This research could extend to investigating the effects that GenAI has on language. Are the LLMs influencing the way that language evolves and if so in what ways? At this point, since GenAI has only been available for a short amount of time, many of the assumptions that we make about its effects on writing and education are tentative. As we study the relationship between LLMs and language and rhetoric, the experience of students and other writers using it, and the way it is affecting human language, we will be in a position to improve our recommendations.

Note

1 In December of 2022, I began experimenting with sending writing along with prompts to OpenAI's API using multi-shot conversations and discovered that it provides relatively useful feedback. By April 2023, I released into the market one of the first generative feedback tools that teachers or students could submit their writing along with prompts and receive feedback from OpenAI using the GPT-3.5 turbo LLM.

Conclusion

GenAI has proven to be significantly disruptive in academic spaces. On the one hand, it presents significant risks, particularly in its potential to disrupt learning and silence student voices. On the other hand, it could be effectively used as a tool that enhances learning. To approach GenAI in productive ways, it is vital to understand that GenAI is a system of interconnected software and hardware components that all have an effect on how the technology generates text. The central component of modern GenAI systems is one or more Transformer model. Systems that perform NLG usually use LLMs—language models that use the Transformer architecture. Transformer models are a significant type of artificial neural network that are both quick to train and have encoded enough features of language that they can be used to effectively generate reasonable simulations of natural language. The LLMs are made up of a lot of linear algebra and weights, and their ability to represent language in a way that computers can run calculations on it is quite impressive indeed. As we make policy, develop curriculum, and interact with students, it is important that we understand how this technology works well enough to accurately identify its strengths, limitations, and the kind of place it can occupy in our educational ecosystems. Further, there are numerous avenues for research, particularly for the field of Writing Studies to better situate GenAI in our theory. Hopefully, here I have made some small contributions to all of this work.

References

Adler-Kassner, L., Clark, I., Robertson, L., Taczak, K., & Yancey, K. B. (2016). Assembling knowledge: The role of threshold concepts in facilitating transfer. In Chris M. Anson and Jessie L. Moore (Eds.), *Critical transitions: Writing and the question of transfer* (pp. 17–47). The WAC Clearinghouse and University Press of Colorado.

Advice | Are We Asking the Wrong Questions about ChatGPT? (2024, April 15). The Chronicle of Higher Education. https://www.chronicle.com/article/are-we-asking-the-wrong-questions-about-chatgpt

Advice | The Case for Slow-Walking Our Use of Generative AI (2024, February 29). The Chronicle of Higher Education. https://www.chronicle.com/article/the-case-for-slow-walking-our-use-of-generative-ai

Amazon EC2 P3 – Ideal for Machine Learning and HPC – AWS (n.d.). Amazon Web Services, Inc. Retrieved February 24, 2024, from https://aws.amazon.com/ec2/instance-types/p3/

Ambrose, S. A., Bridges, M. W., DiPietro, M., Lovett, M. C., & Norman, M. K. (2010). *How learning works: 7 Research-based principles for smart teaching*. Jossy-Bass.

Anderson, S. S. (2023). "Places to stand": Multiple metaphors for framing ChatGPT's corpus. *Computers and Composition, 68*, 102778. https://doi.org/10.1016/j.compcom.2023.102778

Artemeva, N., & Fox, J. (2010). Awareness versus production: Probing students' antecedent genre knowledge. *Journal of Business and Technical Communication, 24*(4), 476–515. https://doi.org/10.1177/1050651910371302

Artificial Intelligence and the Significance Crisis (2024, February 26). The Chronicle of Higher Education. https://www.chronicle.com/article/artificial-intelligence-and-the-significance-crisis

Ayash, N. B. (2020). Critical translation and paratextuality. *Composition Forum, 44*. https://compositionforum.com/issue/44/critical-translation.php

Barad, K. (2007). *Meeting the universe halfway: Quantum physics and the entanglement of matter and meaning*. Duke University Press.

Barbey, A. K. (2018). Network neuroscience theory of human intelligence. *Trends in Cognitive Sciences, 22*(1), 8–20. https://doi.org/10.1016/j.tics.2017.10.001

Bawarshi, A. (2000). The genre function. *College English, 62*(3), 335–360.

Bawarshi, A. (2003). *Genre and the invention of the writer: Reconsidering the place of invention in composition*. Utah State University Press.

Bawarshi, A., & Reiff, M. J. (2010). *Genre: An introduction to history, theory, research, and pedagogy*. Parlor Press; WAC Clearinghouse.

Bazerman, C. (2003). Speech acts, genres, and activity systems: How texts organize activity and people. In *What writing does and how it does it*. Routledge.

Beach, K.D. (1999). Chapter 4: Consequential transitions: A sociocultural expedition beyond transfer in education. *Review of Research in Education*, 24, 101–139.

Beel, J., Gipp, B., Langer, S., & Breitinger, C. (2016). Research-paper recommender systems: A literature survey. *International Journal on Digital Libraries*, *17*(4), 305–338. https://doi.org/10.1007/s00799-015-0156-0

Biesecker, B. A. (1989). Rethinking the rhetorical situation from within the thematic of différance. *Philosophy & Rhetoric*, *22*(2), 110–130.

Bird, S., & Loper, E. (2004). *NLTK : Natural language toolkit* [Computer software]. https://www.nltk.org/index.html

Bitzer, L. F. (1968). The rhetorical situation. *Philosophy & Rhetoric*, *1*(1), 1–14.

Boersma, P., Benders, T., & Seinhorst, K. (2020). Neural network models for phonology and phonetics. *Journal of Language Modelling*, *8*(1), Article 1. https://doi.org/10.15398/jlm.v8i1.224

Bourdieu, P. (1977). *Outline of a theory of practice*. Cambridge University Press.

Brauwers, G., & Frasincar, F. (2023). A general survey on attention mechanisms in deep learning. *IEEE Transactions on Knowledge and Data Engineering*, *35*(4), 3279–3298. https://doi.org/10.1109/TKDE.2021.3126456

Brent, D. (2011). Transfer, transformation, and rhetorical knowledge insights from transfer theory. *Journal of Business and Technical Communication*, *25*(4), 396–420. https://doi.org/10.1177/1050651911410951

Brown, T. B., Mann, B., Ryder, N., Subbiah, M., Kaplan, J., Dhariwal, P., Neelakantan, A., Shyam, P., Sastry, G., Askell, A., Agarwal, S., Herbert-Voss, A., Krueger, G., Henighan, T., Child, R., Ramesh, A., Ziegler, D. M., Wu, J., Winter, C.,…, & Amodei, D. (2020). *Language models are few-shot learners* (arXiv:2005.14165). arXiv. https://arxiv.org/abs/2005.14165

Burchell, J. (2024, May 18). *Lies, damned lies and large language models*. PyConUS 2024, Pittsburgh, PA.

Byrd, A., & Flores, L. (2023). *MLA-CCCC joint task force on writing and AI members*. Modern Language Association and the Confernece on College Composition & Communication. https://aiandwriting.hcommons.org/working-paper-1/

Canagarajah, A. S. (2006). The place of world Englishes in composition: Pluralization continued. *College Composition and Communication*, *57*(4), 586–619.

CCCC. (2018, June 6). Students' right to their own language (with bibliography). *Conference on College Composition and Communication.* https://cccc.ncte.org/cccc/resources/positions/srtolsummary/

ChatGPT. (2024). [Computer software]. OpenAI. https://chatgpt.com

Claude. (2024). [Computer software]. Anthropic. https://claude.ai/chats

Coleman, M. C. (2018). Bots, social capital, and the need for civility. *Journal of Medical Ethics*, *33*(3), 120–132.

Consigny, S. (1974). Rhetoric and its situations. *Philosophy & Rhetoric*, *7*(3), 175–186.

Cress, U., & Kimmerle, J. (2023). Co-constructing knowledge with generative AI tools: Reflections from a CSCL perspective. *International Journal of Computer-Supported Collaborative Learning*, *18*(4), 607–614. https://doi.org/10.1007/s11412-023-09409-w

Cummings, R. E., Monroe, S. M., & Watkins, M. (2024). Generative AI in first-year writing: An early analysis of affordances, limitations, and a framework for the future. *Computers and Composition*, *71*, 102827. https://doi.org/10.1016/j.compcom.2024.102827

Daniel, S., Pacheco, M., Smith, B., Burriss, S., & Hundley, M. (2023). Cultivating writerly virtues: Critical human elements of multimodal writing in the age of artificial intelligence. *Journal of Adolescent & Adult Literacy*, *67*(1), 32–38. https://doi.org/10.1002/jaal.1298

Deemter, K. V., Theune, M., & Krahmer, E. (2005). Real versus template-based natural language generation: A false opposition? *Computational Linguistics*, *31*(1), 15–24. https://doi.org/10.1162/0891201053630291

Devitt, A. J. (2004). *Writing genres*. Southern Illinois University Press.

Devlin, J., Chang, M.-W., Lee, K., & Toutanova, K. (2019). *BERT: Pre-training of deep bidirectional transformers for language understanding* (arXiv:1810.04805). arXiv. https://doi.org/10.48550/arXiv.1810.04805

Dingo, R. A. (2012). *Networking arguments: Rhetoric, transnational feminism, and public policy writing*. University of Pittsburgh Press.

Dobrin, S. I. (2011). *Postcomposition*. SIU Press.

Dobrin, S. I. (2023). *AI and writing*. Broadview Press.

Doheny-Farina, S. (1992). *Rhetoric, innovation, technology: Case studies of technical communication in technology transfers*. MIT Press.

Downs, D., & Wardle, E. (2007). Teaching about writing, righting misconceptions: (Re) envisioning "first-year composition" as "introduction to writing studies." *College Composition and Communication*, *58*(4), 552–584.

Dryer, D. B. (2008). Taking up space: On genre systems as geographies of the possible. *JAC*, *28*(3/4), 503–534.

Dryer, D. B. (2016). Disambiguating uptake: Toward a tactical research agenda on citizens' writing. In A. S. Bawarshi & M. J. Reiff (Eds.), *Genre and the performance of publics* (pp. 60–82). University Press of Colorado.

Dryer, D., & Peckham, I. (2014). Social contexts of writing assessment: Toward an ecological construct of the rater. *WPA, Writing Program Administration*, *38*(1). http://associationdatabase.co/archives/38n1/38n1dryer-peckham.pdf

Durán, J. M., & Jongsma, K. R. (2021). Who is afraid of black box algorithms? On the epistemological and ethical basis of trust in medical AI. *Journal of Medical Ethics*, medethics-2020-106820. https://doi.org/10.1136/medethics-2020-106820

Dyke Ford, J. (2004). Knowledge transfer across disciplines: Tracking rhetorical strategies from a technical communication classroom to an engineering classroom. *IEEE Transactions on Professional Communication*, *47*(4), 301–315.

Edbauer, J. (2005). Unframing models of public distribution: From rhetorical situation to rhetorical ecologies. *Rhetoric Society Quarterly*, *35*(4), 5–24.

Evans-Tokaryk, T. (2014). Academic integrity, remix culture, globalization: A Canadian case study of student and faculty perceptions of plagiarism. *Across the Disciplines*, *11*(2), 1–40. https://doi.org/10.37514/ATD-J.2014.11.2.07

Fiscus, J. M. (2017). Genre, reflection, and multimodality: Capturing uptake in the making. *Composition Forum*, *37*. https://eric.ed.gov/?id=EJ1162165

Fleckenstein, K. S., Spinuzzi, C., Rickly, R. J., & Papper, C. C. (2008). The importance of harmony: An ecological metaphor for writing research. *College Composition and Communication*, *60*(2), 388–419.

Foucault, M. (2002). *Archaeology of knowledge*. Routledge.

Freadman, A. (2002). Chapter 2: Uptake. In R. M. Coe, Lingard, L., & Teslenko, T. (Eds.), *The rhetoric and ideology of genre: Strategies for stability and change* (pp. 39–53). Hampton Press.

Freadman, A. (2012). The traps and trappings of genre theory. *Applied Linguistics, 33*(5), 544–563.

Freire, P. (1968). *Pedagogy of the oppressed.* Seabury Press.

Fui-Hoon Nah, F., Zheng, R., Cai, J., Siau, K., & Chen, L. (2023). Generative AI and ChatGPT: Applications, challenges, and AI-human collaboration. *Journal of Information Technology Case and Application Research, 25*(3), 277–304. https://doi.org/10.1080/15228053.2023.2233814

Gaines, B. R. (1993). An agenda for digital journals: The socio-technical infrastructure of knowledge dissemination. *Journal of Organizational Computing, 3*(2), 135–193. https://doi.org/10.1080/10919399309540199

Gastaldi, J. L. (2021). Why can computers understand natural language? *Philosophy & Technology, 34*(1), 149–214. https://doi.org/10.1007/s13347-020-00393-9

Gemini-1.5—Chat to Supercharge Your Ideas (2024). [Computer software]. Google. https://gemini.google.com

González-Carvajal, S., & Garrido-Merchán, E. C. (2023). Comparing BERT against traditional machine learning text classification. *Journal of Computational and Cognitive Engineering, 2*(4), 352–356. https://doi.org/10.47852/bonviewJCCE3202838

Google (2024a). *Gemini: Ready to debate?* Gemini-1.5. https://g.co/gemini/share/e44241a8f708

Google (2024b). *Gemini—Chat to supercharge your ideas: Genre analysis prompt.* Gemini-1.5. https://g.co/gemini/share/5b2f1eed1a5f

Grynbaum, M. M., & Mac, R. (2023, December 27). The times sues OpenAI and Microsoft over A.I. use of copyrighted work. *The New York Times.* https://www.nytimes.com/2023/12/27/business/media/new-york-times-open-ai-microsoft-lawsuit.html

Hannah, M. A., & Lam, C. (2016). Patterns of dissemination: Examining and documenting practitioner knowledge sharing practices on blogs. *Technical Communication, 63*(4), 328–345.

Hawk, B. (2007). *A counter-history of composition: Toward methodologies of complexity.* University of Pittsburgh Press.

Hunt, A., & Thomas, D. (1999). *The pragmatic programmer: From journeyman to master.* Addison-Wesley Professional.

Jamieson, S., & Howard, R. (2019). Rethinking the relationship between plagiarism and academic integrity. *Revue Internationale Des Technologies En Pédagogie Universitaire / International Journal of Technologies in Higher Education, 16*(2), 69–85. https://doi.org/10.18162/ritpu-2019-v16n2-07

Jarratt, S., Mack, K., Sartor, A., & Watson, S. (2009). Pedagogical memory: Writing, mapping, translating. *WPA: The Journal of Writing Program Administration, 33* (1–2), 46–73.

Jessop, B. (2018). On academic capitalism. *Critical Policy Studies, 12*(1), 104–109. https://doi.org/10.1080/19460171.2017.1403342

Johnson-Eilola, J., & Selber, S. A. (2007). Plagiarism, originality, assemblage. *Computers and Composition, 24*(4), 375–403. https://doi.org/10.1016/j.compcom.2007.08.003

Kocoń, J., Cichecki, I., Kaszyca, O., Kochanek, M., Szydło, D., Baran, J., Bielaniewicz, J., Gruza, M., Janz, A., Kanclerz, K., Kocoń, A., Koptyra, B., Mieleszczenko-Kowszewicz, W., Miłkowski, P., Oleksy, M., Piasecki, M., Radliński, Ł., Wojtasik, K., Woźniak, S., & Kazienko, P. (2023). ChatGPT: Jack of all trades, master of none. *Information Fusion, 99*, 101861. https://doi.org/10.1016/j.inffus.2023.101861

Kotzeva, E., & Anders, B. (2023). Engineering a dialogue with Klara, or ethical invention with generative AI in the writing classroom. *Journal of Academic Writing, 13*(2), Article 2. https://doi.org/10.18552/joaw.v13i2.989

Latour, B. (1999). *Pandora's hope: Essays on the reality of science studies*. Harvard University Press.

Latour, B. (2005). *Reassembling the social: An introduction to actor-network-theory*. Oxford University Press.

Latour, B. (2007). *Reassembling the social: An introduction to actor-network-theory*. Oxford University Press.

Lawrence, A. (2023). *AntConc* (4.2.4) [Computer software]. Waseda University. https://www.laurenceanthony.net/software/antconc/

Liddell, J. (2003). A comprehensive definition of plagiarism. *Community & Junior College Libraries, 11*(3), 43. https://doi.org/10.1300/J107v11n03_07

Machine Learning Simplified (Director) (2017, August 25). *What is a neural network? (C1W1L02)*. https://www.youtube.com/watch?v=n1l-9lIMW7E

Macklin, M. (2019). *Lost in uptake translation: Examining genre negotiations in students' writing performances* [Thesis]. https://digital.lib.washington.edu:443/researchworks/handle/1773/44186

Marche, S. (2022, December 6). The college essay is dead. *The Atlantic*. https://www.theatlantic.com/technology/archive/2022/12/chatgpt-ai-writing-college-student-essays/672371/

Martin, K. (2019). Ethical implications and accountability of algorithms. *Journal of Business Ethics, 160*(4), 835–850. https://doi.org/10.1007/s10551-018-3921-3

McCabe, J. A. (2015). Location, location, location! demonstrating the mnemonic benefit of the method of loci. *Teaching of Psychology, 42*(2), 169–173. https://doi.org/10.1177/0098628315573143

McKenna, S. (2022). Plagiarism and the commodification of knowledge. *Higher Education, 84*(6), 1283–1298. https://doi.org/10.1007/s10734-022-00926-5

Medina, D. (2017). *A transfer subject: Tracing boundary-work and micro-transfer in first-year composition* [PhD Thesis]. https://digital.lib.washington.edu/researchworks/handle/1773/40559

Medina, D. (2018). Writing the boundaries. *Composition Forum, 42*. https://compositionforum.com/issue/42/boundaries.php

Meretzky, S. E. (1995). *Learning ZIL*. https://archive.org/details/Learning_ZIL_Steven_Eric_Meretzky_1995

Merritt, R. (2023, November 15). *What is retrieval-augmented generation aka RAG?* NVIDIA Blog. https://blogs.nvidia.com/blog/what-is-retrieval-augmented-generation/

Mikolov, T., Chen, K., Corrado, G., & Dean, J. (2013, January 16). *Efficient estimation of word representations in vector space*. arXiv.Org. https://arxiv.org/abs/1301.3781v3

Mikolov, T., Yih, W., & Zweig, G. (2013). Linguistic regularities in continuous space word representations. *In Proceedings of the 2013 Conference of the North American Chapter of the Association for Computational Linguistics: Human Language Technologies* (746–751). Atlanta, Georgia. Association for Computational Linguistics.

Miller, C. (2024). *The everyday function of rhetoric*. https://pressbooks.palni.org/writingfordigitalmedia/chapter/the-everyday-function-of-rhetoric/

Miller, C. R. (1984). Genre as rhetorical action. *Quarterly Journal of Speech, 70*, 151–167.

Miller, S. (1991). *Textual carnivals: The politics of composition*. Southern Illinois University Press.

Mok, A. (2023, April 20). *ChatGPT could cost over $700,000 per day to operate. Microsoft is reportedly trying to make it cheaper.* Business Insider. https://www.businessinsider.com/how-much-chatgpt-costs-openai-to-run-estimate-report-2023-4

Motha, S. (2014). *Race, empire, and English language teaching: Creating responsible and ethical anti-racist practice*. Teachers College Press.

Munroe, R. (2017, May 17). *Machine learning*. Xkcd. https://xkcd.com/1838/

Negretti, R. (2012). Metacognition in student academic writing: A longitudinal study of metacognitive awareness and its relation to task perception, self-regulation, and evaluation of performance. *Written Communication, 29*(2), 142–179. https://doi.org/10.1177/0741088312438529

Nelms, G. (2015, July 20). Why plagiarism doesn't bother me at all: A research-based overview of plagiarism as educational opportunity. *Teaching & Learning in Higher Ed.* https://teachingandlearninginhighered.org/2015/07/20/plagiarism-doesnt-bother-me-at-all-research/

New Age Bullshit Generator (n.d.). Sebpearce.Com. Retrieved February 17, 2024, from https://sebpearce.com/bullshit/

Nowacek, R. S. (2011). *Agents of integration: Understanding transfer as a rhetorical act*. Southern Illinois University Press.

Open AI (2024). *ChatGPT: Genre analysis of protest signs.* https://chatgpt.com/share/22a4faac-6fac-451b-bfe5-a8e2ef237e70

Opinion | How Will Artificial Intelligence Change Higher Ed? (2023, May 25). The Chronicle of Higher Education. https://www.chronicle.com/article/how-will-artificial-intelligence-change-higher-ed

Orben, A. (2020). The Sisyphean cycle of technology panics. *Perspectives on Psychological Science*, *15*(5), 1143–1157. https://doi.org/10.1177/1745691620919372

Pennington, J., Socher, R., & Manning, C. (2014). Glove: Global vectors for word representation. *Proceedings of the 2014 Conference on Empirical Methods in Natural Language Processing (EMNLP),* 1532–1543. https://doi.org/10.3115/v1/D14-1162

Pfeifer, R., & Scheier, C. (2001). *Understanding intelligence*. MIT Press.

Quinn, T. P., Jacobs, S., Senadeera, M., Le, V., & Coghlan, S. (2022). The three ghosts of medical AI: Can the black-box present deliver? *Artificial Intelligence in Medicine*, *124*, 102158. https://doi.org/10.1016/j.artmed.2021.102158

Řehůřek, R., & Sojka, P. (2010). Software framework for topic modelling with large corpora. *Proceedings of the LREC 2010 Workshop on New Challenges for NLP Frameworks*, 45–50. https://www.researchgate.net/publication/255820377_Software_Framework_for_Topic_Modelling_with_Large_Corpora. DOI:10.13140/2.1.2393.1847

Reiff, M. J., & Bawarshi, A. (2011). Tracing discursive resources: How students use prior genre knowledge to negotiate new writing contexts in first-year composition. *Written Communication*, *28*(3), 312–337. https://doi.org/10.1177/0741088311410183

Reiter, E. (1996). *Building natural-language generation systems* (arXiv:cmp-lg/9605002). arXiv. https://arxiv.org/abs/cmp-lg/9605002

Rice, J. (2012). *Digital Detroit rhetoric and space in the age of the network*. Southern Illinois University Press. https://search.ebscohost.com/login.aspx?direct=true&scope=site&db=nlebk&db=nlabk&AN=466185

Rounsaville, A. (2012). Selecting genres for transfer: The role of uptake in students' antecedent genre knowledge. *Composition Forum, 26*. https://compositionforum.com/issue/26/selecting-genres-uptake.php

Rounsaville, A. (2014). Situating transnational genre knowledge: A genre trajectory analysis of one student's personal and academic writing. *Written Communication, 31*(3), 332–364.

Sánchez Martín, C., Hirsu, L., Gonzales, L., & Alvarez, S. P. (2019). Pedagogies of digital composing through a translingual approach. *Computers and Composition, 52*, 142–157. https://doi.org/10.1016/j.compcom.2019.02.007

Scrapy Developers (2024). *Scrapy | a fast and powerful scraping and web crawling framework* (2.11) [Computer software]. https://scrapy.org/

Sebpearce/Bullshit: Generates New Age Nonsense on the Fly (n.d.). Retrieved February 17, 2024, from https://github.com/sebpearce/bullshit

Selby, C., & Woollard, J. (2013). *Computational thinking: The developing definition* [Monograph]. University of Southampton (E-prints). https://eprints.soton.ac.uk/356481/

Shipka, J. (2011). *Toward a composition made whole*. University of Pittsburgh Press.

Shute, V. J., Sun, C., & Asbell-Clarke, J. (2017). Demystifying computational thinking. *Educational Research Review, 22*, 142–158. https://doi.org/10.1016/j.edurev.2017.09.003

Smith, A. (2016). *Heteropatriarchy and the three pillars of white supremacy: Rethinking women of color organizing.* https://doi.org/10.1215/9780822373445-007

Sønderby, C. K., Espeholt, L., Heek, J., Dehghani, M., Oliver, A., Salimans, T., Agrawal, S., Hickey, J., & Kalchbrenner, N. (2020). *MetNet: A neural weather model for precipitation forecasting* (arXiv:2003.12140). arXiv. https://doi.org/10.48550/arXiv.2003.12140

Sparck Jones, K. (1972). A statistical interpretation of term specificity and its application in retrieval. *Journal of Documentation, 28*(1), 11–21. https://doi.org/10.1108/eb026526

Spinuzzi, C. (2003). *Tracing genres through organizations: A sociocultural approach to information design.* MIT Press.

Spinuzzi, C. (2008). *Network: Theorizing knowledge work in telecommunications.* Cambridge University Press.

Stanfordnlp/GloVe: Software in C and Data Files for the Popular GloVe Model for Distributed Word Representations, a.k.a. Word Vectors or Embeddings (2015, October 24). https://github.com/stanfordnlp/GloVe/tree/master

Steen, M. (2015). Upon opening the black box and finding it full: Exploring the ethics in design practices. *Science, Technology, & Human Values, 40*(3), 389–420. https://doi.org/10.1177/0162243914547645

Sternberg, R. J. (2000). *Handbook of intelligence*. Cambridge University Press.

Stevenson, M., & Phakiti, A. (2014). The effects of computer-generated feedback on the quality of writing. *Assessing Writing, 19*, 51–65. https://doi.org/10.1016/j.asw.2013.11.007

Swales, J. M. (1990). *Genre analysis: English in academic and research settings.* Cambridge University Press.

Tachino, T. (2012). Theorizing uptake and knowledge mobilization: A case for intermediary genre. *Written Communication, 29*(4), 455–476.

Taylor, B. L. (2023, November 15). *Long hours and low wages: The human labour powering AI's development.* The Conversation. https://theconversation.com/long-hours-and-low-wages-the-human-labour-powering-ais-development-217038

Tenney, I., Das, D., & Pavlick, E. (2019). *BERT rediscovers the classical NLP pipeline* (arXiv:1905.05950). arXiv. https://doi.org/10.48550/arXiv.1905.05950

The Value of AI in Today's Classrooms (2024, June 11). Walton Family Foundation. https://www.waltonfamilyfoundation.org/learning/the-value-of-ai-in-todays-classrooms

TOEFL Test Takers (n.d.). ETS. Retrieved June 15, 2024, from https://www.ets.org/toefl.html

Use of Generative Artificial Intelligence in Courses: Artificial Intelligence at Northwestern – Northwestern University (n.d.). Retrieved June 14, 2024, from https://ai.northwestern.edu/education/use-of-generative-artificial-intelligence-in-courses.html

Vaswani, A., Shazeer, N., Parmar, N., Uszkoreit, J., Jones, L., Gomez, A. N., Kaiser, Ł., & Polosukhin, I. (2017). *Attention is all you need.*

Vaswani, A., Shazeer, N., Parmar, N., Uszkoreit, J., Jones, L., Gomez, A. N., & Polosukhin, I. (2023). Attention Is All You Need. *arXiv [Cs.CL]*. Retrieved from http://arxiv.org/abs/1706.03762

Vatz, R. E. (1973). The myth of the rhetorical situation. *Philosophy & Rhetoric*, *6*(3), 154–161.

Wallace, R. (1995). *A.L.I.C.E.: Artificial linguistic internet computer entity* [Computer software].

Wang, P. (2019). On defining artificial intelligence. *Journal of Artificial General Intelligence*, *10*(2), 1–37. https://doi.org/10.2478/jagi-2019-0002

Wardle, E. (2009). "Mutt genres" and the goal of FYC: Can we help students write the genres of the university? *College Composition and Communication*, *60*(4), 765–789.

Weizenbaum, J. (1966). ELIZA—A computer program for the study of natural language communication between man and machine. *Communications of the ACM*, *9*(1), 36–45. https://doi.org/10.1145/365153.365168

What Is Intellectual Property (IP)? (n.d.). Retrieved May 12, 2024, from https://www.wipo.int/about-ip/en/index.html

Wing, J. M. (2006). Viewpoint: Computational thinking. *Communications of the ACM*, *49*(3), 33–35.

Winner, L. (1993). Upon opening the black box and finding it empty: Social constructivism and the philosophy of technology. *Science, Technology, & Human Values*, *18*(3), 362–378. https://doi.org/10.1177/016224399301800306

Wolfram, S. (2023, February 14). *What is ChatGPT doing ... and why does it work?* https://writings.stephenwolfram.com/2023/02/what-is-chatgpt-doing-and-why-does-it-work/

Word Completion for Text Documents (n.d.). Retrieved May 27, 2024, from https://help.libreoffice.org/latest/en-US/text/swriter/guide/word_completion.html

WTO | Intellectual Property (TRIPS)—What Are Intellectual Property Rights? (n.d.). Retrieved May 12, 2024, from https://www.wto.org/english/tratop_e/trips_e/intel1_e.htm

Zork Source Code (1977). [ZIL]. Massachusetts institute of technology. https://github.com/MITDDC/zork (Original work published 2020).

Index

For Product Safety Concerns and Information please contact our EU representative GPSR@taylorandfrancis.com
Taylor & Francis Verlag GmbH, Kaufingerstraße 24, 80331 München, Germany

www.ingramcontent.com/pod-product-compliance
Lightning Source LLC
LaVergne TN
LVHW010937110826
845149LV00013B/2635